Shahid Mahmood

ECOLOGY OF CRITICALLY ENDANGERED ORIENTAL WHITE-BACKED VULTURE

D1826121

Shahid Mahmood

ECOLOGY OF CRITICALLY ENDANGERED ORIENTAL WHITE-BACKED VULTURE

Catastrophic decline of GYPS BENGALENSIS

VDM Verlag Dr. Müller

Impressum/Imprint (nur für Deutschland/ only for Germany)

Bibliografische Information der Deutschen Nationalbibliothek: Die Deutsche Nationalbibliothek verzeichnet diese Publikation in der Deutschen Nationalbibliografie; detaillierte bibliografische Daten sind im Internet über http://dnb.d-nb.de abrufbar.

Alle in diesem Buch genannten Marken und Produktnamen unterliegen warenzeichen-, marken- oder patentrechtlichem Schutz bzw. sind Warenzeichen oder eingetragene Warenzeichen der jeweiligen Inhaber. Die Wiedergabe von Marken, Produktnamen, Gebrauchsnamen, Handelsnamen, Warenbezeichnungen u.s.w. in diesem Werk berechtigt auch ohne besondere Kennzeichnung nicht zu der Annahme, dass solche Namen im Sinne der Warenzeichen- und Markenschutzgesetzgebung als frei zu betrachten wären und daher von jedermann benutzt werden dürften.

Coverbild: www.purestockx.com

Verlag: VDM Verlag Dr. Müller Aktiengesellschaft & Co. KG
Dudweiler Landstr. 99, 66123 Saarbrücken, Deutschland
Telefon +49 681 9100-698, Telefax +49 681 9100-988, Email: info@vdm-verlag.de

Herstellung in Deutschland:
Schaltungsdienst Lange o.H.G., Berlin
Books on Demand GmbH, Norderstedt
Reha GmbH, Saarbrücken
Amazon Distribution GmbH, Leipzig
ISBN: 978-3-639-23765-8

Imprint (only for USA, GB)

Bibliographic information published by the Deutsche Nationalbibliothek: The Deutsche Nationalbibliothek lists this publication in the Deutsche Nationalbibliografie; detailed bibliographic data are available in the Internet at http://dnb.d-nb.de .

Any brand names and product names mentioned in this book are subject to trademark, brand or patent protection and are trademarks or registered trademarks of their respective holders. The use of brand names, product names, common names, trade names, product descriptions etc. even without a particular marking in this works is in no way to be construed to mean that such names may be regarded as unrestricted in respect of trademark and brand protection legislation and could thus be used by anyone.

Cover image: www.purestockx.com

Publisher:
VDM Verlag Dr. Müller Aktiengesellschaft & Co. KG
Dudweiler Landstr. 99, 66123 Saarbrücken, Germany
Phone +49 681 9100-698, Fax +49 681 9100-988, Email: info@vdm-publishing.com

Printed in the U.S.A.
Printed in the U.K. by (see last page)
ISBN: 978-3-639-23765-8

Dedicated to

Those who are my every thing

My consolation in sorrow,

My hope in misery,

My strength in weakness,

Whom they have sourced of,

Love, Mercy, Sympathy,

Forgiveness, Pleasure,

Enjoyment and Inspiration

My Parents
&
My Wife

ACKNOWLEDGEMENT

I would bow my head before **Almighty Allah**. The omnipotent, the omnipresent, the merciful, the most gracious, the compassionate, the beneficent, who is the entire source of all knowledge and wisdom endowed to mankind and who blessed me with the ability to do this work and **Hazrat Muhammad** (Peace Be Upon Him) who gave us the spirit to learn. It is the blessing of **Allah** that has enabled me to make my efforts a success.

I would like to take this opportunity to convey my gratitude and appreciation to my respected supervisor **Professor, Dr. Aleem Ahmed Khan**, Institute of Pure and Applied Biology, Bahauddin Zakariya University Multan, without whose constant help, deep interest and vigilant guidance, the completion of this research work would not have been possible. I am really indebted to him for his accommodative attitude, patience, sympathetic behavior and administrative measure in providing me all the facilities required to complete this piece of work.

I pay my profound thanks to **Dr. Martin Gilbert, Dr. Munir Z. Virani, Dr. Patrick Benson, Dr. Richard T. Watson** and **Dr. J. Lindsay Oaks**, **The Peregrine Fund, America** and **Ornithological Society of Pakistan** for their Valuable guidance and financial support to complete this piece of work.

I would like to express my gratitude to **Professor, Dr Muhammad Faheem Malik** Head Department of Environmental Sciences, University of Gujrat, Gujrat whose sincere efforts made me proficient to reach at this stage.

It is necessary to pay thanks to all my research fellows especially to **Shakeel Ahmed, Muhammad Jamshed Iqbal Chaudhery, Ahmed Ali, Muhammad Arshad,** and **Abdul Razaq** for their co-operation during the course of my work.

Humble thanks to my father in law **Rana Shahbaz Khan** and my **mother in law** for their buttress, encouragement and love.

Above all, I would feel it necessary to express my exclusive love and adoration to my father **Haji Muhammad Sharif Shami** and my **mother**. I want to thank you for your unconditional support. I am honoured to have you as my parents. Thank you for giving me a chance to prove and improve myself through all my walks of life.

I wish to extend my utmost gratitude to my sisters, my brothers **Muhammad Sabir, Muhammad Khalid, Kashif Hussain, Muhammad Naqqash,** and brothers in law **Abdul Wahid, Muhammad Asif, Muhammad Azeem, Abaid Ur Rehman, Rana Amir Shahbaz** and **Aziz Rabbani**. Thank you all for believing in me. I Hope that you will be able to fulfill your dreams.

I would also like to express my heartfelt gratitude and love to my **Wife** and to my little angel **Muhammad Subhan Shahid** for their never ending love, care and moral support for me.

Finally, I thank all those who assisted, encouraged and supported me during this research, be assured that **Allah Almighty** will bless you all for the contributions you made.

Shahid Mahmood
Lecturer,
Department of Zoology
University of Gujrat,
Gujrat, Pakistan

CONTENTS

Description	Page Number

Description	Page Number

CHAPTER -06

CHAPTER -07

LIST OF TABLES

LIST OF FIGURES

LIST OF FIGURES

CHAPTER -01

INTRODUCTION

There are 15 species of Old World Vultures *Aegypiinae*. Out of which eight species are reported from the Indian subcontinent (Ali and Ripley 1983). Of which seven species have been observed in Pakistan. These are King Vulture (*Srcogyps calvus*), Cinereous Vulture (*Aegypius monachus*), Egyptian Vulture (*Neophron percnopterus*), Eurasian Griffon (*Gyps fulvus*), Himalayan Griffon (*Gyps himalayensis*), Long-billed Vulture (*Gyps indicus*) and White-backed Vulture (*Gyps bengalensis*) (Roberts 1991).

The Oriental White-backed Vulture *Gyps bengalensis* has been described as the commonest vulture in the Indian subcontinent with a range extending to South Vietnam and the Malay Peninsula (Ali and Ripley 1968). The species is resident in Pakistan and is "widely distributed throughout the provinces of Punjab, Sind and the broader valleys of North West Frontier Province" (Roberts 1991). The species "prefers cultivated tracts with scattered trees and a high human population, being attracted to larger towns and cities, where slaughterhouses and refuse tips offer more opportunity for obtaining food" (Roberts 1991).

The White-backed Vulture occurs in south-east Iran, south-east Afghanistan, Eastern Pakistan, India, Nepal, Bhutan, Bangladesh and Myanmar (Birdlife International 2001). It is absent from the north Himalayas, and parts of north-east, north-west and south-west India, and Sri Lanka. It was also formerly widely distributed in South-East Asia in Thailand, Malaysia, Laos, Cambodia, Vietnam and southern China (south and west Yunnan), but it is now almost extinct in this region, although relict populations survive in a few remote areas. There is one record of a vagrant in Borneo (MacKinnon and Phillips 1993).

In Pakistan this species is found throughout the Indus plain, being widely distributed through Punjab, Sindh, and the broader valleys of the North West Frontier Province, but absent from the drier mountainous areas and from the Himalayas, and less often in extensive desert tracts such as the Cholistan and Thar deserts, and west of the Hab river on the borders of Sindh and Balochistan (Roberts 1991).

Where the White-backed Vulture is common, in its preferred habitat of wooded savanna, it often outnumbers all other vultures together in the ratio of 2:1 or even 3:1. This figure is derived from counts at carcasses, and taking the counts at face value; however, the ratio assumes that all vultures in an area will gather at a carcass, but this is not so. Nevertheless, the species can certainly give the impression of being remarkably numerous: 200 can visit the carcass of an Impala, while up to 2000 birds might gather around the scene of an Elephant cull (Mundy *et al.* 1992).

A second way of trying to gauge their abundance is by counting flying and perched birds while one travels along in a car (Mundy *et al.* 1992).Without a doubt, the best way of estimating numbers of birds is to count breeding pairs at nests. A bird `density' figure is available from Zululand in north-east South Africa. Here, before the arrival of a severe cyclone, 106 pairs nested in about 1000 sq. km, which is nearly 0.3 birds per sq. km, although the vultures no doubt also foraged in surrounding tribal areas. After cyclone Demoina caused a heavy flood in January 1984, which removed many riverine trees, 242 nests were counted in the two game reserves (Mundy *et al* 1992).

The Oriental white-backed Vulture (*Gyps bengalensis*), have declined catastrophically throughout the Indian subcontinent (Prakash 1999, Cunningham 2000). Since the mid-1990s, there has been a growing realization that populations of vultures of the genus *Gyps,* and particularly of the white-backed vulture *(Gyps bengalensis),* have declined (Prakash 1999). Studies by the Bombay Natural History Society (BNHS) have provided convincing evidence that normal ecological factors, such as food supply and habitat availability, are not constraining the Indian vulture populations. A disease is suspected to be the cause of the population declines (Oaks 2000, Cunningham 2000). An additional concern is the spread of the disease, if it exists, to vulture populations in other parts of the world (Trivedi 2001). The most significant histopathological lesions seen in post-mortem examination of the dead birds in Pakistan, and India are those of avian visceral gout (Oaks 2000).

White-backed Vulture (*Gyps bengalensis*) has been upgraded to Critical because it has suffered an extremely rapid population decline, particularly across the Indian subcontinent, probably as a result of disease compounded by poisoning, pesticide use and changes in the processing of dead livestock (BirdLife International 2001).

Declines in Oriental White-backed Vulture populations were first reported in India by Prakash (1999) in Keoladeo National Park, Rajasthan. Maximum vulture counts within the park declined by 96% from 1985/86 (max. 1800) to 1998/99 (max. 86). Numbers of active nests showed a comparable decline of 95%, with 353 nests located in 1987-88 and just 20 in 1998/99. Breeding success was shown to have fallen from 82% of eggs laid in 1985/86 to 0% in 1997/98 and 1998/99 (Prakash and Rahmani 1999). Numbers of G. *indicus* sighted in Keoladeo National Park had also declined, from a maximum count of 816 birds in 1985/86 to only 25 in 1998/99 (Prakash 1999). Reports from numerous sources presented in Birdlife International (2001) suggested that the decline in populations of G. *bengalensis* and G. *indicus* were more widespread, extending across much of peninsular India. Observations suggest that the recently described Slender-billed Vulture *Gyps*

2

tenuirostris may also have been affected (Birdlife International 2001). A reduction in the population of other sympatric species of Old World vulture, the Egyptian Vulture *Neophron percnopterus* and Red-headed Vulture *Sarcogyps calvus,* have not been described, suggesting that the declines may be restricted to the genus *Gyps*.

A 96% decline in the population was observed over the past decade. During 1985-86 the highest population of 2000 vultures (density = 69/km^2) was recorded whereas a maximum of only 86 vultures was recorded during 1998-99 (density = 3/km^2) (Prakash 1989).The nesting population of vultures also crashed by 95% during this decade. For instance, 363 pairs (nest density of 12.2 nests/km^2) were recorded nesting during 1987-88 (Prakash 1989) but only 150 nests were recorded during 1996-97, 25 nests during 1997-98 and just 20 (nest density = 0.7 nests/km^2) in 1998-99). A sharp decline in the breeding success of this vulture was also recorded during the past decade. The nesting success was recorded as 82% of eggs laid (n = 244) in 1985-86 but was nil during 1997-98 (n = 25) and 1998-99 (n = 20).

A wide range of hypotheses have been proposed to explain a decline of this magnitude and rate. These include: a reduction in food availability, loss of suitable nesting habitat, pesticide intoxication, deliberate poisoning, emerging infectious disease, and calcium deficiency (Prakash 1999). Investigations into the cause of the declines have been underway since 1999 (Oaks *et al* 2001, Cunningham *et al* 2001, Pain 2001) and have largely suggested the involvement of an infectious disease or pesticide intoxication. Visceral gout was reported to be a "common finding" of vulture post mortem examination in India (Pain 2001). Avian visceral gout should not be regarded as a disease entity, but as a clinical sign of any severe renal dysfunction (Lumeij 1994) and may be considered an end stage in a broad range of infectious and non-infectious disease processes. As yet no single underlying factor has been identified to account for the visceral gout that has been found in the vulture population.

Declines for similar reasons appear to have occurred in at least parts of Bangladesh since the 1970s. By contrast, in the Indian subcontinent, this species was common until recently, with high population densities in many urban areas maintained by abundant supplies of livestock carcasses. Indeed, it was once considered "possibly the most abundant large bird of prey in the world" (Birdlife International 2001). However, in the late 1990s populations in India and Nepal crashed, with 95-100% declines reported. These rapid declines are now also occurring in Pakistan.

Documentation of bird mortality is difficult, as most carcasses go undiscovered. Of those found, there are biases toward man-induced factors (Newton 1976). The proximate and ultimate causes of death may differ and are often difficult to assess, leading to questionable conclusions. Though many causes of mortality have been suggested for Cape Vultures Gyps *coprotheres* (Benson and Dobbs 1984, Tarboton and Allan 1984, Mundy *et al.* 1997), little is known of the

3

yearly variation in these factors or their relative importance to the total population or particular sub-populations.

In the Delhi-Agra,Bharatpur belt, nearly 18,000 vultures are estimated to have died over the last decade (Satheesan 2000).

In Pakistan in July 2000, 165 White backed vultures were seen along 15 km of canal bank in south Punjab (D. G. Khan, Muzaffargarh and Layyah districts), none showing head-drooping behaviour, and locals reporting no unusual mortality patterns; over 500 individuals observed on cattle carcasses in this area in June 2000. However, in August 2000 a total of 1,366 White-backed Vultures were seen along 1,809 km in Punjab province, and 16.6% of 175 individuals observed at close quarters at seven sites showed head-drooping behaviour indicative of disease (Khan *et al.* 2001). One of these sites (Changa Manga plantation) held a population of more than 500 birds. Dying individuals and local reports of vulture deaths (in the last 2-3 years during the hot months of April-July) were noted at Head Islam (near Hasil Pur), and specifically for White-backed Vulture at Dinga Nalla (near Ghazi Ghat), and in Lal Sohanra National Park (Khan *et al.* 2001). Overall the proportion of individual's head-drooping, and number of deaths reported, were highest in areas near the border with Rajasthan, India, and lower in the Indus River areas (Rahmani and Prakash 2000). At Kundian forest, Mianwali district, "large-scale deaths" of vultures were found in April May 2000; although birds were reported to "have stopped dying" by August and no sick-seeming individuals were observed in that month. A similar epidemic in 1992 was reported by local people (BirdLife International 2001).

Little is known about the post-nestling (post-fledging) dependence period in the *Falconiformes* and especially the causes of mortality during this period (Newton 1979).More specifically Mundy (1982) found that, while juvenile Cape Vultures(*Gyps coprotheres*) still receive food from their parents, after leaving the nest they are subject to a high mortality.

Autopsies on vultures performed so far have found "degenerative changes in the urinary tubules" in the kidneys, and whitish deposits presumed to be urates present in the heart, liver, kidney and spleen; no evidence of bacterial infection was found, so organ samples were taken for culture in an effort to detect viruses, but no results are available yet (Risebrough 1999). The progressive accumulation of uric acid (as found in the human condition commonly termed gout) may therefore explain the pattern of reproductive failure and chronic condition eventually resulting in death, as observed in vultures in Keoladeo National Park (Risebrough 1999). Seven more dead vultures from the area "had died of an infectious disease, probably a virus. The actual cause of death appeared to be dehydration caused by enteritis" (Prakash 2000). All autopsies carried out so far on vultures found dead in the wild have shown symptoms of acute enteritis, degeneration of kidney cells, and extensive visceral gout, whilst vultures which were captured as sick individuals that died in captivity did not show visceral gout (Rahmani and Prakash 2000). A wildlife pathologist from the Zoological Society of London concluded that the available evidence strongly indicates an infectious disease to be responsible (Rahmani and Prakash 2000). Results from attempts by the Poultry Diagnostic and Research Centre, Pune, to isolate viruses in vulture tissue have indicated the presence of a viral pathogen, and electron microscopy work at the National Virology Laboratory, Pune, has confirmed the presence of viral particles in vulture tissue samples (Rahmani and Prakash 2000). However, it must still be stressed that such findings do not necessarily mean that the disease agent is a virus (Virani *et al.* 2001).

Postmortem analysis of *G. bengalensis*, *G.indicus*, and one *G.himalayensis* from India (Cunningham *et al.* 2001) and *Gyps bengalensis* from Pakistan (Oaks *et al.* 2001, Gilbert *et al.* 2002) identified renal and visceral gout (crystallization of uric acid in the tissues) in the majority of birds found dead, and enteritis in high proportion of the birds from India (Cunningham *et al.* 2001).

The presence of visceral gout in tissues of dead birds from both countries supports the hypothesis that the same mortality factor is responsible for all deaths. Although renal gout is often attributed to kidney disease, in theses cases gout was acute (occurring only a few hours before death), suggesting that this condition is a consequences of primary disease and not a disease it self (Cunningham *et al.* 2001).

Visceral gout and enteritis are non specific lesions and could result from , for example a contaminant result or an infectious disease process. Histological analysis of tissues from Indian birds, however, found higher then expected proportion of vultures' with inflammation of blood vessel walls and proliferation in the brain of glial cells (inflammatory cells specific to the central nervous system (Cunningham *et al.* 2001).

Head drooping occurred in vultures while they are perched either on trees or on the ground: their heads and necks drooped almost to the point of touching their feet, giving a sickly and lethargic impression. Mundy *et al.* (1992) occasionally observed similar behaviour in African White-backed Vultures *Gyps africanus* and describe it as dozing. However, in India, an unusually high frequency of head drooping has been observed in vultures even during cool conditions (Prakash 1999). This head drooping could be symptomatic of lethargy induced by an infectious disease or a physical result of a disease itself.

Symptoms of individuals of the two species dying in Keoladeo National Park, India, were consistent with the neck repeatedly slumping slowly down before being jerked back up (Prakash 1999). Individuals remained in this sick condition for more than 30 days before falling off their perch and dying entangled in branches beneath or on the ground below, although some sick individuals could fly short distances and even feed their young (Prakash 1999).

Head-drooping behaviour was noted in 17% of individuals of White-backed Vultures during surveys by BNHS in April June 2000 in north and central India (Prakash 2000). Head-drooping behaviour in apparently sick individuals was observed at numerous other sites in India, and also in Nepal and Pakistan in 2000 (Khan *et al*. 2000).

The aim was to measure mortality rates among vultures and determines the causes of mortality to understand the factors responsible for the population collapse in Punjab, Pakistan.

Aims and Objectives

The major objectives of the study are,

➢ To determine the total vulture population and its decline at Toawala and Dholewala.

➢ To ascertain the population density/kilometer area.

➢ To determines the patterns of mortality in *Gyps bengalensis*.

➢ To analyses the characteristic of head drooping behaviour with different ecological parameters.

CHAPTER -02

STUDY AREA

In relation to the catastrophic decline of Oriental White-backed Vulture (Prakash 199), a hypothesis was developed to as certain if there is any problem with the habitat, food etc. of the species in Pakistan. It has been mostly observed that many of the bird's species reach to the critical endangered level due to change or alteration or loss of habitat (Bird Life 2001). Keeping the hypothesis in view a study was devised in Punjab province. In this regard during the initial survey our field studies were conducted in Punjab Province of Pakistan at two main sites (Toawala and Dholewala). Toawala (N: 30°40'26", E: 70°55'11"), is located in Khanewal District. It is a representative population from the Southern central Punjab and is about 5 Km east to the River Chanab and comprises elevated canal banks lined with mature Sheesham trees *Dalbergia sisoo* that provide a linear distribution of suitable vulture nest sites. Dholewala (30° 32' 48" N, 70° 52' 43" E) in Muzaffargar and Layyah Districts, It is a representative site of the Southern Punjab with ample population of Oriental White-backed Vulture. It is located east of the Indus River and comprises elevated canal banks lined with mature Sheesham trees *Dalbergia sisoo* that provide a linear distribution of suitable vulture nest sites.

The elevation of Oriental White-backed Vultures nesting sites at Dholewala and Toawala are 600 feet from the sea level. The region in the vicinity of the nesting site experienced a dry subtropical climate with an annual rain fall of 200 – 450 mm, and a relative humidity ranging from 25 – 85 %.The average minimum temperature in January is 4.5 – 5.5 c°, and the average maximum in June is 42 – 45 c°. The minimum temperature recorded is 0 c° and maximum is 49 c° (www.wetlands.org).

The land-use is mainly agricultural with cotton and wheat as the principal crops. Total area of colonies was 39.50 Km and 64.1 Km at Toawala and Dholewala Colony respectively. In two intensive transects both at Toawala and Dholewala total area of intensive transects was 7.67 Km and 5.0 Km at Toawala and Dholewala sites respectively.

Abundant herds of domestic livestock occur in all sites. Live stock rearing together with agriculture is the basis of economy in these both Study areas at Toawala and Dholewala. Live stock is comprised of sheep, cattle, buffalos, goats, horses and camels. Among wild species we found the wild boar (*Sus scrofa*).Carnivore representative species are the fox (*Vulpes vulpes*) the dogs (*Canis familiaris*) .

Main vegetation at both study sites Dholewala (DW) and Toawala (TW) is sheesham (*Dalbergia sisso*), Mango (*Mangiera indica*), Jaman (*Szyginum cumuni*), Neem (*Azadirachta indica*), Bakain (*Melia azedarachta*), Date palm (*Phoenix dactyliera*) and Safeda (*Eucalyptus camaldulensis*). Other natural vegetation includes *Acacia nilotica, Prosopis cineraria, Pisum arvense, Salsola baryosma, Cynodon dactylon, Eleusine compressa* and *Panicum antidotale*,(www.wetlands.org).

Other raptors sighted at both study sites during vultures observations included *Gyps fulvus* (Eurasian Griffon), *Accipiter nisus, Linn* (Sparrow Hawk), *Aquila vindhiana, Frankl* (Tawny Eagle), *Astur badius, Falco tinnunculus* (Common Kestril), *Circus melano*leucus *frost* (Pied Harrier*), Haliaster Indus* (Brahmini kite), *Pernis ptilorhynchus* (Honey buzzard), *Falco chicquera* (Red necked falcon), *Milvus migranus* (Black kite) *Ethen brama* (Spotted owl), *Hieraaetus fasciatus* (Bonelli's Eagle) and *Elanus caeruleus* (Black shouldered Kite) etc.

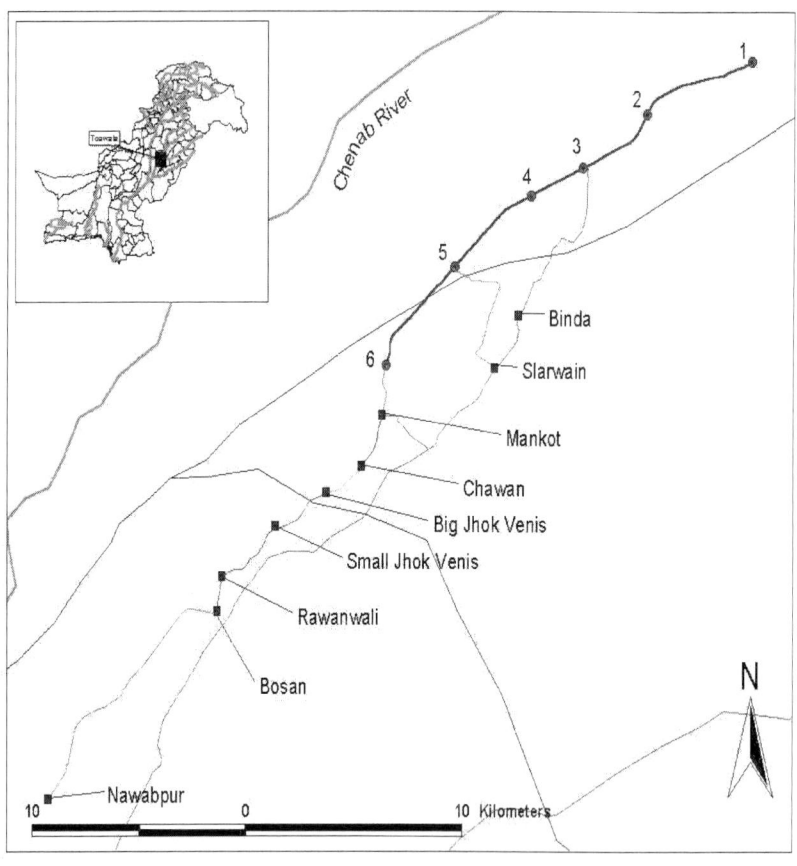

Fig 1: Colony of Oriental White-backed Vulture at Toawala study site in Punjab Province, Pakistan.

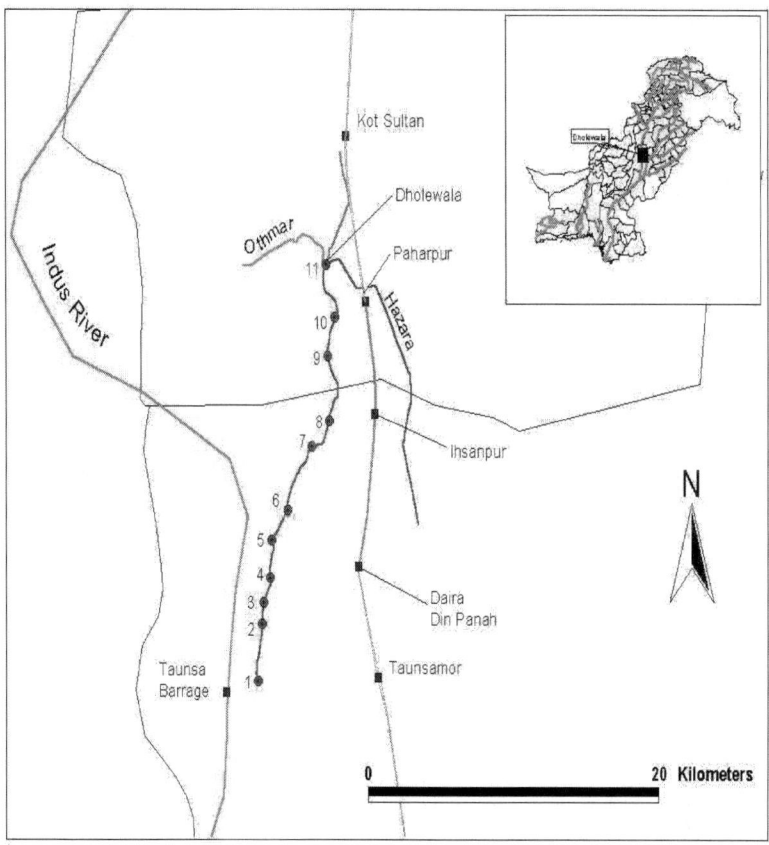

Fig 2: Colony of Oriental White-backed Vulture at Dholewala study site in Punjab Province, Pakistan.

CHAPTER -03

LITERATURE REVIEW

Birds are the most familiar of all creatures. While the most animals are shy, avoid humans and are therefore seen only occasionally. But birds are quite different. Common birds can be seen and heard almost every where. Till recently the vultures were common. These large birds come under the order *Falconiformes* of family *Cathartidae*, which covers New World vultures, and the family *Accipitridae* covers Old World vultures. The New World vultures are distributed from southern Canada to Tierradel Fuego up to the Falkland Island, and the Old World vultures are distributed in Africa, Europe and Asia. Although the Old World vultures strongly resemble their new world counterparts in appearance and behavior, but are not related (Chhangani and Mohnot 2001). The Subfamily *Aegypinae* contains 15 species of Old World vultures .There are 15 species of Old World Vultures *Aegypiinae*, which are not systematically related to the New World Vultures *Cathartidae*. The curious Palm Nut Vulture or Vulturine Fish Eagle *Gypohierax angolensis* seems to provide a transition in habits, and to some extent in anatomy, from the sea eagles to the Old World Vultures. It may well be related to both, though this cannot be taken as proved.. Of the latter, *Gypohierax* most resembles the Egyptian Vulture *Neophron percnopterus*, which is black and white like some sea eagles, but differs in many ways from *Gypohierax*. The unique Bearded Vulture or Lammergeier *Gypaetus barbatus* has some similarities to *Neophron*. The two genera are associated, but perhaps *Gypaetus*, deserves a major group of its own. The other genera of Old World' vultures (*Necrosyrtes, Gyps, Aegypius, Torgos, Sacrogyps, Triginiceps*) are all closely related to each other. The Old World vultures occur over much of the Old World, excluding Australasia. Vultures observed in many parts of Europe (Spain, France, Italy, Greece), Africa (Kenya, Tanzania, Zimbabwe, Botswana, South Africa, Namibia, Angola) and Asia (India, Yemen, Israel, Caucasus, Mongolia)unpublished material accumulated by the World Working Group on Birds of Prey (WWGBP).

During the 1990s nine vulture species were recorded in South Africa, seven as breeding species. Seven of these nine species are listed in *The Eskom Red Data Book of Birds of South Africa, Lesotho and Swaziland* (Barnes 2000). Two species are not listed; the Ruppell's Griffon is a vagrant to South Africa and the Palm-nut Vulture *(rare* in the previous red data book; Brooke 1984) only marginally occurs in this country (Harrison *et al.* 1997). The Egyptian Vulture is classified as *extinct,* the Bearded Vulture as *endangered,* and the other four species are listed in the *vulnerable* category. Various factors are responsible for the current conservation status of vultures (Mundy *et al.*1992) and because of their precarious status;

scavenging raptors are recognized as one of the most threatened guilds of birds in South Africa (Barnes 2000).

If two (or more) different species populations require a common resource that is potentially limited and actually becomes so; they are said to be in competition for it (Gause 1934). When such a situation arises one of the species would eliminate all others directly or through competitive exclusion resulting in ecological isolation. Therefore congeneric species are known to be isolated from each other by range, habitat or feeding habits (Lack 1971). A world review of birds with reference to coexistence of congeneric species by Lack (1971) shows only one exception where two species *'(Calidris melanotos & C. alpinus)* with identical feeding habits share the same range ¨and habitat for a short period, of 10 to 12 weeks, but that is attributed to temporary .supply of superabundant food. Subsequent workers (Vijayan 1975 and Houston 1975) too have added fresh data to strengthen the theory of ecological isolation originally put forward by 'Gause and further developed by others.

The species was once described as fairly common over the whole Thailand (Gyldenstolpe1916), from the plains to 1,600 m (Lekagul and Round 1991) although avoiding the more densely wooded parts (Gyldenstolpe1920). It was "common everywhere" in Thailand, although "not extending south of Taiping, in the Malay Peninsula" (Robinson and Kloss 1921-1924), and it was regarded as "common" in 1919 in Phuket (Robinson and Kloss 1921-1924, Medway and Wells 1976, Wells 1999). It nested in the early part of the twentieth century around Chiang Mai in northern Thailand (Gyldenstolpe 1916). it was "very common on the plain", and "large numbers" roosted on Doi Sutep to at least 550 m "as well as in tall trees throughout the city" (Deignan 1931, 1936). Vulture populations around Chiang Mai declined rapidly around the early 1960s, apparently as a result of poisoning from strychnine-laced meat put out during a campaign to reduce the number of stray dogs around the city in 1960 (Cheke 1972). Even though this species was still fairly common in the 1960s, within a decade it had become rare, and in 1985 it was estimated to be on the verge of extinction (Round and Chantrasmi 1985).In Malaysia this species was once a widespread resident occurring along the full length of the peninsula until the early part of the twentieth century, but it is now very local and sparse (Wells 1999).

This species was once abundant in suitable habitat throughout Laos (Thewlis *et al.* 1998). In Tranninh it was found commonly (Delacour and Jabouille 1927), and even in the mid-twentieth century it was still abundant there and also in Savannakhet province (David-Beaulieu 1944, 1949-1950). In Champasak and Attapu provinces it was common, but outnumbered by the Red-headed Vulture (Engelbach 1932). However, it is now restricted to the southern part of Champasak and Attapu provinces (Thewlis *et al.*1998), with populations well below carrying capacity and of "miserably low" breeding output and it may soon become extinct (Thewlis *et al.*1998).The species was once abundant in suitable habitat throughout the Cambodia, but by the 1960s it was regarded as uncommon (Thomas 1964).

Historically, this species was abundant in suitable habitat in central and south Vietnam.The species was being very common on the Pleiku plateau (David-Beaulieu 1939), "common" in South Annam (Delacour *et al.* 1928), and "the commonest vulture" in south Vietnam, being "often seen in large groups" (Wildash 1968). It is now almost extinct (Thewlis *et al.* 1998), with only one recent record in Dak Lak province (Le Xuan Canh *et al.* 1997), but none was recorded during recent surveys in Dak Lak province in February-May 1998 (Brickle *et al.* 1998), and none was found during surveys in December 1989-March 1990 at sites throughout Vietnam (Robson *et al.* 1990).

The general abundance, until recently, of this species and other vultures in India compared to other countries in its range has been attributed to the human population's traditional avoidance of eating beef, so cattle carcasses are abandoned and scavenged by vultures and other species (Rahmani 1998). The smaller populations of vultures in several Asian countries, and perhaps in southern India in Kerala and parts of Tamil Nadu, may be because beef is more commonly eaten in these regions, so fewer cattle carcasses are available to the birds. However, in the late 1990s population declines and disease symptoms were noted throughout India, from Rajasthan in the west to Assam in the east, south to at least Maharashtra, Karnataka and Andhra Pradesh; in the southern states of Kerala and Tamil Nadu, declines from historically smaller populations may have occurred prior to the recent crashes, probably as a result of a combination of other threats (Rahmani 1998).

A small population of Cinereous Vulture *Aegypius monachus* breeds in the country but the majority of the population is migratory. The latter species is seen as far south as Kutch in Gujarat during the winter months. The Bearded Vulture *Gypaetus barbatus* and Himalayan Griffon *Gyps himalayensis* are typical mountain species and breed in the upper reaches of Himalayan and Trans-Himalayan regions. The young Himalayan Griffons winter in the Himalayan foothills and <u>sometimes</u> as far south as Kutch in Gujarat (Samant *et al.* 1995). The Eurasian Griffon *Gyps fulvus* breeds in Himalayas but is seen throughout the Indian plains to Deccan plateau during winter. The White-backed Vulture *Gyps bengalensis,* Long-billed *Gyps indicus* and Egyptian *Neophron percnopterus* are the most common species of vultures and are known to breed from foothills of Himalayas south till Kanyakumari. The White-backed Vultures are most numerous and are seen in hundreds on large ungulate carcasses over most parts of north, central and east India and up to Karnataka in the south India. The Long-billed is largely a cliff nester and is common at most locations within its distribution except in the southern states of Tamil Nadu and Kerala. The Egyptian Vulture is common over most of its distribution except in northeastern and south India. The

King Vulture *Sarcogyps calvus which* was never very numerous and had its distribution from the foothills of Himalayas to the entire subcontinent, has become quite rare in western and eastern India but is holding on in other parts of its distribution (Samant *et al.* 1995).

In Nepal the White-backed Vulture is found throughout the lowlands up to about 1,000 m, less frequently up to 1,800 m, and rarely up to 3,050 m (Inskipp and Inskipp 1991). This species has only recently been recorded from Bhutan, and it is now known from: Deothang, 1994-1998 (Bishop 1999).

This species is resident in the west and formerly also occurred elsewhere in the Myanmar, where it's current status unknown (Robson 2000). This species was once resident throughout the Thailand, but it is now close to extinction. Records are known from: Bangkok, undated (Williamson 1914). This species was once a widespread resident occurring along the full length of peninsular Malaysia, but now only occurs very rarely as a non-breeding visitor (Wells 1999). This species was once found throughout Laos, but it is now restricted to southern parts of Champasak and Attapu provinces in the south of the country (Thewlis *et al.* 1998). This species was recently described as resident throughout the Cambodia (Robson 2000). In Vietnam this species was formerly resident in South Annam and Cochin China (Robson 2000), but it is now almost extinct. This species is widely distributed in India, from the Himalayas west to Srinager, east to Arunachal Pradesh, Assam and the north-east hill states, south to the southern Western Ghats in Kerala and Tamil Nadu (Bates and Lowther 1952).

It is a resident bird and nests about 50 km south west of the Park. The birds are however seen in the Park throughout the year. Their numbers start increasing from November and the population reaches a peak in March-April. The population of Long-billed Vultures appears to have decreased by over 97% during the past decade. A maximum of 816 birds (density = 28 birds/km2) were recorded during 1985-86 (Prakash 1989) but only 25 birds (0.9 birds/km^2) were recorded during 1998-99.

Mortality factors documented at the Kransberg, the largest Cape Vulture breeding colony, where reproductive activities were monitored and carcasses collected as part of a long-term study of this bird's biology (Benson & Dobbs 1984; Dobbs & Benson 1984, 1984; Tarboton & Benson 1988; Benson *et al.*. 1990).

More than one sixth of the world's population lives in the south Asian countries of India, Nepal and Pakistan. These countries have more races, more languages, more religions and more social groups (tribes and castes) than any comparable corner of the globe. This incredible diversity of peoples is one of the region's great fascinations. Their enlightened and benevolent attitudes towards wildlife have undoubtedly helped to conserve the rich natural heritage of the region, which supports as many as 13% of the world's birds. Of these, 141 species are endemic, a total comprising over 10% of the region's avifauna. Nine species of vultures occur in the region of which five belong to the genus *Gyps*. No single group of wild birds has been more beneficial to the people of the region than *Gyps* vultures. Cattle are -considered sacred by the region's predominantly Hindu and Buddhist communities, and Muslims will not consume livestock that have died of natural causes. Vultures consume the numerous carcasses of cattle and other livestock, stripping them to the bone and helping rid the landscape of a potential source of disease-causing organisms. This natural disposal phenomenon has become an integral part of the region's human dominated ecology over thousands of years creating a mutualistic association between man and bird. Among Zoroastrian communities living in the Indian subcontinent, vultures have helped maintain a 3,000-year old tradition by consuming human corpses at `Towers of Silence'. This tradition, along with other benefits that vultures provide is now threatened. A new menace threatens to wipe out the three species of *Gyps* vultures (Oriental White-backed *Gyps bengalenis,* Cliff *Gyps indicus* and Slender-billed *Gyps tenuirostris)* that, occur in the region. It has decimated *Gyps* vulture populations throughout India and Nepal and threatens the last remaining large populations of vultures in the Indus River Valley of Pakistan (Prakash 2000). *Gyps* vultures in south-east Asia are already virtually extinct, though possibly from some other, unrelated cause. The cause of mortality remains unidentified but is\ suspected to be an infectious disease (Riseborough 1999, Cunningham 2000).The proximate causes of breeding failure of birds in the Keoladeo National Park are also most plausibly explained by a disease factor (Risebrough 1999).

Dead crows *Corvus* have also been found in Keoladeo National Park, and as this species also commonly feeds at carcasses, these deaths may also have been caused by the same disease factor (Prakash 1999). This has potentially alarming human health transmission by a mosquito vector of a virus of African origin resulted in the infection and death of both wild birds (including crows) and a number of people in the New York City region during the summer of 1999 (Risebrough 1999).

Although the toxic effect of pesticides has been implicated in population declines of a number of raptor species (Ratcliff 1967, Hickey and Anderson 1968) and this has been suggested as an explanation of the population crashes in Gyps vultures in the Indian region (Rahmani 1998, Ghatak 1999, Prakash 1999), the evidence does not suggest that this factor can be responsible for the catastrophic recent declines, although it may represent a low-level threat.

It has been rumoured that a "cultural offensive" against vultures has been launched in Myanmar, involving dumped pesticides on baits, but this cannot be confirmed. In Peninsular

Malaysia, sodium arsenite was the prevalent herbicide in rubber estates at the time that vulture populations were present but dwindling, was "a notorious killer of wandering village stock", and hence might have led indirectly to vulture deaths. In India, instances of strychnine poisoning were reported as long ago as 1888 in Assam (Hume 1888). Predators such as jackal *Canis aureus*, wolf C. lupus, leopard *Panthera pardus*, tiger and lion *P. Leo*, which may attack domestic livestock, are occasionally targeted by poisoning of carcasses (Prakash 1999). Poachers also poison wild animals to facilitate removal of hides, antlers and horns, etc. (Satheesan 2000). Furthermore, poison is occasionally administered to domestic cattle in order to facilitate the removal of the hide: the rodenticide zinc phosphate has been used for this purpose around Keoladeo National Park, India (Prakash 1999). Some instances occurred in Gir forest, Gujarat, India (Grubh 1974), and in Kutch, Gujarat, in May 1999 locals reported "every vulture in and around Anaimalai Hills", Tamil Nadu, being killed by poisoned carcasses put out to kill cattle-marauding leopards between 1960 and 1980 (Kannan 1993); and about 40 died at a poisoned carcass in Rangpur district, Bangladesh, in May 1981 (Sarker 1983). However, these are relatively rare events, and mortality of vultures from this source does not seem to be a major risk and it cannot be countenanced as the cause of the recent collapse in India's population (Prakash 1999); furthermore, the poisons used for this purpose are acutely toxic and would be expected to kill the birds rapidly, with lower sublethal doses inducing sickness followed by recovery (Risebrough 1999), and a chronic condition that deteriorates over time would not be expected unless the toxic effects were reinforced through repeated exposure (Risebrough 1999). This view is nevertheless disputed (Satheesan 2000).

In Bangladesh, there is a local superstition that hanging a vulture's head around the neck of a cow will reduce maggot infestations in wounds during the rainy season, and hence some hunting of vultures occurs for this purpose (Sarker 1983).

Several measures have been taken to eliminate populations near to airports to reduce the hazard to aviation, including direct killing of vultures, modernising of nearby slaughterhouses, and banning of carcass dumping (Satheesan 1999, Rahmani and Prakash 2000). More than a decade ago population declines as a result of these measures had been noted in White-backed and Indian Vultures in a number of urban areas (Satheesan 2000).

The effect of the factors described above may have been exacerbated by low levels of genetic diversity (Prakash 1999) as has been found in the Cape Vulture (Wyk *et al.* 1993). No studies have investigated this in the White-backed Vulture or in any of the other Indian species (Prakash 1999), but given the large contiguous range and dispersive abilities of vultures, this seems unlikely.

The sudden loss of vultures from the Indian subcontinent, where carcass disposal is unsophisticated and inefficient, raises concerns over increased human and livestock health hazards. The loss of vultures from the ecosystem will most likely have negative consequences for ecosystem structure and function, resulting in ecological imbalances. If an infectious disease is decimating vultures in the Indian subcontinent the possibility exists that it could spread to other vulture species in the Middle East, Europe and Africa. In Africa, avian scavengers alone account for 65% of wild ungulate consumption in the savannah grasslands (Houston 1979). A similar crash of vultures in Africa would have dire consequences for ecosystem structure and function.

Improved hygiene and carcass disposal methods, as well as concerted efforts by aviation authorities to reduce air strikes from vulture collisions at major airports, may explain declines around large cities. In agricultural areas where vultures forage, poisoning from exposure to toxins as a result of an increase in the use of pesticides and other poisons may be responsible for some level of population decline (Roberts 1991, Grimmet 1998, Rahmani 1999). However, these reasons seem unlikely to explain the sub-continent wide decline in vulture numbers. This decline has been genus-specific, with only *Gyps* vultures apparently affected (Prakash 2000; Riseborough 1999); poisoning would be expected to affect all the species exposed to it. There is also no evidence for use of a new, specific type of toxin on a regional scale (Riseborough 1999). However, investigations into the possible role of poisoning are ongoing. Other environmental factors such as temperature extremes, possibly resulting from global warming, and contaminated water have also been suggested as compounding factors contributing to large-scale vulture mortalities. While the principal cause of vulture deaths remains unknown, there is evidence that an infectious disease, possibly viral, could be responsible (Cunningham 2000). Necropsies conducted in India in March 2000 found that dead vultures showed signs of avian visceral gout (the deposition of uric acid in the viscera) and histology showed perivascular lymphocytic cuffing (PVLC) consistent with inflammation from infection (Cunningham 2000, Prakash 2000). Vultures in India were described as dying with visceral gout, evidence of enteritis and PVLC following protracted periods of head drooping behaviour.

Declines in these species were first noticed by Dr Vibhu Prakash, Principal Scientist at BNHS, who had been monitoring vultures at the Keoladeo National Park, Bharatpur, Rajasthan since the mid1980s (Prakash 1999). Peak numbers (March/April) of White-backed Vultures declined dramatically from a maximum of 1800 in 85/86 to 86 in 98/99. Similar declines were found in the non-resident Long-billed Vultures that nest 50 km away. Breeding success declined over this period to zero, and sick and dead adult and young birds were found. Birds exhibited lethargy and a head

17

drooping almost to their feet for protracted periods (up to 4 or 5 weeks) before dropping out of trees. Food availability remained constant over this period, suggesting that poisoning (from pesticides or other chemicals) or disease may have been factors in the declines (Prakash 1999).

Eurasian (*Gyps fulvus*) and Himalayan (*Gyps himalayensis*) Griffons are migratory and there is continuous distribution of these birds throughout Africa and Europe. Traditionally, Eurasian Griffons have wintered in India. Their range stretches westwards to Europe, and in North Eastern Africa overlaps with that of Rueppell's Griffon (*Gyps rueppellii*), who's range in turn overlaps in eastern Africa with that of the African white-backed Vulture (*Gyps africanus*). The exception was, when the Eurasian Griffons were sighted in Kenya for the first time (Trivedi 2001). In conclusion, the Eurasian Griffons, if affected, are the pathway of the infectious agent to the Middle East, Europe and Africa (Risebrough & Virani, 2000).

In general, little is known about the post-nestling (post-fledging) dependence period in the Falconiformes and especially the causes of mortality during this period (Newton 1979). This is certainly true for the Old World vultures (Mundy 1982), where only in the case of the Cape Vulture *Gyps coprotheres,* is some information available on the post-nestling dependence period (Robertson 1985).

Vultures receive food only from their own parents and only at their own nest sites (Robertson, 1985); it is imperative that they are able to recognize the sites themselves and also their parents at these sites during this critical period. This recognition might be expected for the survival of a colonially living species in which post-nestling parental care is prolonged. The need for, and ability of, individuals of certain colonial species to be able to identify each other (here adult-to-adult and also adult to offspring and *vice versa) is* reviewed by Falls (1982).

Young Cape Vultures can recognize their parents flying nearby; Robertson (1985) records that juveniles indicated parental recognition by starting to beg before the parent bird had landed. Our observations and those of Robertson (1985) show that fledglings alight at their own nest sites even when their parents are absent; they are later fed there by their parents. However, fledglings were more often stimulated to revisit their own nest sites by the presence there of one or both parents (Robertson 1985). In addition adults recognize their own offspring, since breeding adults will reject begging fledglings from other sites (Mundy 1982; Robertson 1985).

At the present stage of the evolutionary development of this species the ability of fledglings to recognize their nest sites, and also their parents, must already be highly developed as non-recognition, apparently leading to certain death, would have been severely selected against. Also some vultures were rejected by their own parents after they had left the nest. Such action could relate to the parent offspring conflict, as discussed by Trivers (1974), in which selection between the inclusive fitness of the juvenile and the reproductive fitness of the parent occurs (Wilson 1975). It is thus a question of whether the parents continue feeding their offspring or whether they do not and rather start initiating the following season's breeding cycle. The condition (indicated by total body mass) of the newly flown bird may then be important; for example Jarvis (1974) found that in the Cape Gannet *Sula capensis* chicks with a high mass at a nest departure have a higher survival rate than those with a low mass.

If the fledglings had been repulsed by their parents it is noteworthy that, even though they were hungry, they did not persist in trying to obtain food from their parents, but rather begged from other adults. Why, if at all, were they so easily repulsed? Robertson (1985). In the absence of findings other then vasculities ,gliosis is generally associated with viral infection.

All the species of vultures are scavengers and are commensal of man. Their occurrence in extremely high density is attributed to the availability of abundant food supply due to the primitive method of carcass disposal. All of them have successfully exploited the vast food resource created by man by extensive dairy and poultry farming. The carcasses of livestock form the principle food for vulture and are now mostly dependent on man's activity for the food. With the dwindling population of wild animals, food has become limited in the wilderness areas (Prakash & Rahmani 2000).

The species feeds almost entirely on carrion, mainly by scavenging at rubbish dumps and slaughterhouses, and by searching for dead animals by soaring on thermals (Grimmett *et al.* 1998). In Keoladeo National Park, India, it feeds on carcasses of frail and old cattle abandoned by villagers in the park, and also on individuals that die after getting trapped in the mud of drying marshes (Prakash 1999). In the Sundarbans, Bangladesh, it feeds on the carcasses of cattle but also of wild boar, deer, monkeys, and occasionally tigers *Panthera tigris*, which are found floating in rivers and channels (Sarker 1987). On average, one adult vulture eats approximately 1 kg meat per day (Sarker 1987). A mixed-species vulture flock was reported to eat clean a carcass of a freshly dead bullock within 40 minutes (Ali and Ripley 1968-1998), and flocks of 200-400 were regularly seen cleaning carcasses in 15-20 minutes (Satheesan 1989). In Bombay, White-backed Vultures flew up to 25 km from roost sites to feed at a carcass-processing plant at Korakendra (Singh *et al.* 1996), but Roberts (1991) estimated that individuals may travel well in excess of 300 km per day in search of food, given their effective use of thermal currents. In cities in India, high populations of vultures were formerly maintained by abundant food supplies at slaughterhouses, bone mills (where vultures are utilized to pick the bones clean before crushing, for production of tallow and glue) (Satheesan 1989), carcass-processing factories, tanneries, and garbage dumps (Grubh 1983). In Bombay, White-backed and (less commonly) Indian Vultures play an important cultural role in the religious Parsi community, who place their dead in the "Towers of Silence" on Malabar hill for the vultures to dispose of (Grubh 1983, Houston 1990, Satheesan 1998).

In South East Asia the decline in vulture populations, which occurred largely during the early to mid-twentieth century, was attributed to "shooting, live capture for display, and the great reduction in carrion" due to improvements in hygiene, with other possible factors being pesticides and deaths on roads (P. D. Round 1998). In Thailand, the main causes of the population decline were believed to be a great reduction in the availability of carrion prey, and disturbance of nest sites (Round and Chantrasmi 1985), although strychnine poisoning from laced carcasses (aimed at reducing the stray dog population) was held to be responsible for vulture declines around Chiang Mai (Cheke 1972). Improvements in animal husbandry, and hence a reduction in the supply of carcasses, was regarded as the main cause of population declines in the Malay Peninsula (Wells 1999). Suitable breeding habitat remains widespread in Laos and adjacent countries at least, so this factor is unlikely to have played an important part in the significant population declines there. In this region, the relatively slow decline of vulture populations, and the fact that Red-headed Vulture and most other large birds (e.g. adjutant storks *Leptoptilos*, Black Kite *Milvus migrans* and Brahminy Kite *Haliastur indus* in Laos at least) have exhibited comparable declines, suggests that the presumed disease which is affecting vultures in the Indian subcontinent is unlikely to be implicated (P. D. Round 1998). Very little is known about the ecology of vultures in this region, or about the main factor limiting the remaining populations, but supply of carcasses and disturbance by humans are likely to be amongst the most important. In Myanmar, declines in this species were attributed to poisoning by insecticides (Sayer and Usan Han 1983).In the Indian subcontinent a number of threats have contributed to population declines, particularly the modernization of slaughterhouses, poisoning of carcasses, and reduction in nesting trees around cities.

In the Sundarbans, Bangladesh, "a drastic decrease in the numbers of livestock" was reported to have affected numbers of White-backed Vultures (Sarker 1987). Elsewhere in Indian subcontinent, however, changes in the food supply have not occurred within the short time-period necessary to explain the rapid vulture population declines (Risebrough 1999). For instance, at Keoladeo National Park, India, during 1985-1986, vultures were found on every carcass encountered, with a mean of 80 vultures per carcass (Prakash 1999). By 1998-1999 only 8% of 100 carcasses had vultures feeding on them, with a mean of 19 individuals (Prakash 1999). The supply of carcasses has not changed (20-25 per month), and there has been no change in the method of carcass disposal by villagers around the park: cattle carcasses are still thrown out into the open after being skinned (Prakash 1999). In 1999 a survey conducted around the park in Bharatpur district found that approximately 2,500 cattle die every month, which was estimated to provide sufficient food to sustain a population of 13,000 vultures (Rahmani and Prakash 2000). Villagers keep fewer cattle now in northern India owing to the liberalisation of the economy, and in many areas villagers are shifting to higher-quality breeds that are well looked after and suffer lower mortality, so fewer carcasses are available to vultures (Rahmani 1998). However, this trend is not widespread in the "cow-belt" states of northern India where vultures are commonest (Rahmani 1998). Surveys by BNHS in north and central India in April June 2000 found only 5% of 192 livestock carcasses had vultures in attendance (Prakash 2000). It is noteworthy that both White-backed Vulture individuals on which autopsies have been carried out contained abundant fat in the abdominal cavity, suggesting that they had not died of malnutrition (Risebrough 1999).

The phenomenon of five or six species of vultures feeding together at large carcasses has often been observed in different parts of Africa. This variety of scavenger species has led o much speculation about their ecological separation (Chapin 1932, Petrides 1959, Attwell 1963, ruuk 1967, Houston 1975). Only Kruuk (1967) and Houston (1975) have investigated this phenomenon in detail. Kruuk divided the six common species of vultures found n East Africa into three different pairs, according to the structure of their bills and their related feeding behaviour He then suggested that each pair of vulture species was separated ecologically from the other pairs, primarily by eating different parts (tissues) of the carcass - the two griffon vulture species eating soft tissues, the Lappet-faced and White-headed Vultures eating skin and tendons, and the Hooded and Egyptian *(Neophron percnopterus)* Vultures eating small scraps and pecking between bones.

Changes in land-use, poisoning and food availability are the major factors which will limit vulture numbers in the future. Land-use changes and declining food availability are linked, and ongoing, and impact on all wildlife populations. These impacts will only be stopped if human populations are controlled. Poisoning has been a major influence throughout the world in the decline of *Gyps* and other vultures (Brown 1977; Schenk 1977; Benson & Dobbs 1984; van Jaarsveld 1987). Without control of poisoning, land-use changes and food availability become insignificant as there will not be any birds to worry about. Though efforts have been made to curb poisoning, it continues. The challenge in the future will be to find innovative ways to educate and involve the farming community in vulture conservation and protection. Food resources for scavengers will continue to decline as human populations increase and farming methods become more sophisticated. It is likely that vulture numbers will decline to levels which can be supported by local conditions. This is acceptable if the remaining population is healthy and self-sustaining, however with lower numbers any poisoning incidents are likely to have a relatively greater impact on surviving populations, making it vital to control this source of mortality.

Drooping and dead vultures of all age classes have also been observed in Pakistan and Nepal, suggesting that the problem is prevalent on a wider scale. If this is the case, then *Gyps* species in Europe, Middle Eastand Africa may soon be affected too populations in India are continuous with others all the way to the southern tip of Africa.In India organochloride pesticides are used extensively in agriculture (Prakash 1999), and this is also true in other countries in the range of the declining Gyps vultures (Ghatak 1999). For example, around Keoladeo National Park, India, farmers extensively used organochloride compounds such as aldrin, dieldrin, endosulfan and heptachlor; moreover, DDT has been banned for use in agriculture in India, but it is still extensively applied after being diverted from the national malaria control programme (Prakash 1999). High levels of DDT and HCH pesticides were found in tissue samples from cattle and pig carcasses collected from areas surrounding the park, and these two pesticides (plus, in some cases, dieldrin) were found in carcass tissue samples collected in Rajasthan, Uttar Pradesh and Delhi (Ghatak 1999). Lethal levels of DDE (the main metabolite of DDT), aldrin and dieldrin were detected in the tissues of Sarus Crane (*Grus Antigone*) and Ring Dove (*Streptopelia decaocto*) in Keoladeo National Park (Vijayan 1991).

However, there is no direct evidence for significant pesticide levels in the small number of vulture tissue samples collected from the park so far (Prakash 1999, Rahmani and Prakash 2000).Deliberate poisoning of carcasses (e.g. with strychnine) normally to rid a neighborhood of scavenging predatory mammals-has been identified as a major source of mortality in other raptors (Dobbs and Benson 1984), and cases have been reported for vultures in India (Satheesan 2000) and Thailand (Cheke 1972).Lead poisoning from gunshot used by poachers has also been suggested as a possible risk to vultures (Satheesan 2000).

Virtually no hunting or trapping of vultures occurs in the Thar Desert, Rajasthan (Rahmani 1996). However, vulture eggs, chicks and adults were caught for food with nets, nooses and using bare hands, in a number of areas including: Guntur and Prakasam districts, Andhra Pradesh, until 1980 (K. M. Rao 1992, Satheesan 2000).

Aircraft strikes Mortality due to collision with aircraft has been a long-recognized hazard at airports in India (Ali and Grubh 1984, Thiollay 2000), particularly at Delhi, Bombay, Calcutta, Hyderabad, Madras, Trivandrum and Bangalore (Satheesan 1989), with at least 15 military aircraft and several human lives being lost due to vulture strikes during 1980-1994 (Satheesan 2000), and vultures being involved in 39% of 265 aircraft strikes recorded by Ali and Grubh (1984).

CHAPTER -04

MATERIALS AND METHODS

Intensive field-based ecological studies was designed to measure population and mortality rates, and help identifying the causes of mortality through field necropsies and the laboratory analysis of tissue samples collected for pathogens and contaminants. The main effort had been to collect fresh tissue from vultures that had died with the symptoms of visceral gout. Comprehensive protocols had been developed for assessing numbers and behaviour of vultures.

Binocular was used for identification of different age class group during the data collection process of head drooping and vulture count. While HOBO logger (a computer chip) was placed in both sites for regular temperature recording analysis. Necropsy kit was used for the necropsy of the every fresh dead bird. The "GPS" (Global positioning system) was also used to record the waypoint of each dead bird. For this GPS apparatus (G.P.S. 12 Channel Garmin) was used, which showed longitude and latitude of the area. The simple correlation analysis was used to determine the relationship between variables. Minitab program was used for analyses of the data.

The sampling protocols were designed to document and monitor the following aspects of vulture biology:

Vulture count

Surveys were carried out from June 2001 to June 2002 at various locations through out the Punjab Prvince. Where the regular surveys were carried out at two colonies (Toawala & Dholewala) of Punjab Province. The road transect method was followed (Fuller and Mosher 1981) in both sites. Numbers of vultures were also estimated at carcass and garbage dumps.

Overall data was collected on individual observation and population and structure such as species, age class, and the presence of other terrestrial scavengers.

Toawala and Dholewala were subdivided into transects. Each of these transects started and ended at a 'transect point' (01, 02, 03 etc). Thus transects were divided into 01-02, 02-03, 03-04 etc. In cases where counts had been carried outside the regular transects, site locations were mentioned. Vulture Count Sheet had been intended to record information during site visits to colonies, outside the main study transects. However, sheets were occasionally used to record counts of birds within the study transect, particularly during the earlier part of the season (Appendix 1a).

Start/End GPS (N)/ (E) Degrees minutes and seconds had been calculated in decimals. Distance covered was recorded in km. As GPS systems were not available during the earlier portion of the season, many of these distances represent estimates, odometer readings, or measurement based on the marker stones at the field sites.

Number of nests seen with no birds in attendance (these may still be active, as none were physically checked during this study) called empty nests. Occupied nests were number of nests with birds in attendance.

Total area survey for the census was found 39.50 Km and 64.1 Km at Toawala and Dholewala respectively and the density of vultures per kilometer area documented was counted.

For the sub sample in intensive study transects both at Toawala and Dholewala Colony, 7.67 Km and 5.1 Km transects were survey on weekly basis and vultures were counted on weekly basis. All aspects were studied as described above (Appendix 1b).

Head-drooping behaviour

Head drooping occurred in vultures while they are perched either on trees or on the ground: their heads and necks drooped almost to the point of touching their feet, giving a sickly and lethargic impression.

Data was collected on proportions of drooping to non drooping vulture flocks at various sites and plotted on a temporal scale. The frequency of head-drooping was also correlated with parameters such as age class, drooping intensity, perching substrate, size of crop, temperature and orientation to sun (Appendix 3).

Head drooping data had been collected by making an assessment of birds sitting in 'groups'. The word 'group' simply refers to the numbers of birds visible from a single position.

Number of droopers refers to the number of birds within the 'group' adopting a drooping posture (full or half), each vulture was assigned a number. Those birds seen to be drooping are recorded first. Hobo logger was used for recording of temperature at both sites. Where temperature was not taken, the mean temperature of the same month of either previous year or next year was taken for correlation analysis of droopers with temperature. Pearson correlation test was used for correlation analysis. Number of non droopers was also correlated with the crop status.

Mortality and morbidity

Colonies were visited, as often, as possible (preferably daily) and vulture carcasses collected for further analysis. Age class, species, estimated time of death; position and state of carcass were also recorded to provide a crude rate of mortality at colonies (Appendix 2a).

Dholewala (Dw) and Toawala (Tw) had been subdivided into transects. Each of these transects starts and ends at a 'transect point' (01, 02, 03 etc). Thus transects were divided into 01-02,

02-03, 03-04 etc. Note had been made of dead birds found close to study sites. The study site had been divided by marker stones placed every 300m, these had been used to divide study transects up into more manageable portions. As for example, TW bird found on 09/05/01 in TWTP 03-04 was apparently found between marks 95 and 96. Where nest number was indicated it had been recorded in parentheses e.g. (001E).GPS was taken as Degrees, minutes and seconds had been calculated in decimals.

Data was presented to measure rate and identify patterns of mortality of Oriental White-backed Vulture in Punjab Province, Pakistan between mid December 2000 to the end of July 2001 and from October 2001 to August 2002.

Data was also presented on the occurrence of visceral gout in relation to body condition and age class of vulture. A method of estimating mortality rate within the breeding populations was described to draw spatial and temporal comparisons in this study. An estimation of annual mortality rate was calculated by dividing the number of dead vultures (DV) located within the study plots during the observation period (OP in days) by the number of breeding individuals (based on the number of active nests (AN) at the start of the observation period) and extrapolating this over a 365 day period; using the formula:

$$\frac{DV/(AN \times 2)}{OP} \times 365$$

Several vultures were post mortem during which the body condition of these birds and the opportunity was taken to examine their digestive tracts.

Mortality was measured by methodically collecting and removing dead vultures from beneath known nests throughout the breeding season. Study sites were searched for dead vultures at both main sites. Detailed necropsies were performed on those dead vultures where autolysis was not yet advanced. Where possible, the cause of death was determined from gross examination, and samples were collected for histological, microbiological and toxicological analysis that was not part of this study. Observation of white uric acid precipitate over the surface or within the parenchyma of the liver, the kidneys, the lungs, within the pericardium, joints (femoro-tibial, tarso-metatarsal, shoulder), or any one of these was considered sufficient to indicate the presence of visceral gout (Appendix 2b).

Houston (1976) used a system for scoring body condition of dead vultures by assessing the extent of fat reserves in the omentum, mesentery and subcutaneous tissue. Houston's scoring technique was simplified in this study and a bird was considered to possess omental fat reserves if deposits of fat formed a definite mass that obscured the intestines when abdominal musculature was resected. Birds with less extensive fat deposits were recorded as negative for omental fat.

The rate of decomposition varied greatly throughout the observation period, being significantly higher during the hot months of April, May and June. Evidence of uric acid accumulation was seen to persist several days after decomposition prevented detailed post mortem examination. In cases where it was not possible to conduct a more detailed post mortem, the skin, musculature and body wall were incised to expose the thoracic and abdomenal viscera. An assessment of visceral gout and omental fat was made in these birds as described above, and where possible the sex of birds was determined. No assessment of articular gout was made in these birds.

The age class of dead vultures was recorded by using plumage characteristics for adult (full adult plumage), sub adult (1 year to adult), juvenile (fledging to 1 year), and nestling. Time since death was estimated in periods of 0 -1 day, 2 - 7 days, 8 - 30 days, and > 31 days. The visual assessment of decomposition and the frequency of visits to the site, ensured that time since death were categorized into these broad classes with reasonable accuracy. Vulture carcasses were removed from the site, or where this was not possible, carcasses were buried, to avoid double counting (Appendix 2c).

Tissues collected during post mortem were transferred to Normal Buffer formaline.This formaline was prepared by one of the following methods.

Formaline Preparation

Key features are that it must have the correct pH, and that it should be a 10% solution of 37-40% formaldehyde. Make and work with these solutions either in a hood or a very well ventilated area, and wear gloves and eye protection.

TO MAKE 1 LITER (1000 ML) – METHOD # 1

Formaldehyde 37-40% stock solution	100 ml
Sodium Phosphate, Monobasic	4 grams
Sodium Phosphate, Dibasic	6.5 grams

Added 500 ml distilled water, and adjust pH to 7.2
Added distilled water to a final volume of 1000 ml (about 350-400 ml)

TO MAKE 1 LITER (1000 ML) – METHOD # 2

Formaldehyde 37-40% stock solution	100 ml
10 X concentrated PBS (phosphate buffered saline)	100 ml

Added 500 ml distilled water, and adjust pH to 7.2 with concentrated hydrochloric acid and sodium hydroxide as needed.
Added distilled water to a final volume of 1000 ml (about 300 ml).

Postmortem Procedure

Material for Necropsy

- Dry shipper/-80 freezer for freezing samples immediately
- Gloves and Whirl packs
- Scalpel holder and blades
- Forceps (rat tooth and flat)
- Large scissors (for cracking rib cage) and Pliers for cracking skull
- Surgical scissors
- Razor blades
- Vernier Calipers and Weighing Scale
- Metal meter rule
- 2 Cutting boards
- Fine point permanent marker (for labeling sample tubes)
- Aluminium foil
- 2-3 tubes containing alcohol for any parasites
- 20-25 small (2ml) orange top tubes for frozen samples
- 2-4 Wide top bottles containing buffered formaline

Procedure

I did the following procedure for necropsy/Autopsy

- Weigh the bird.
- Take tarsus measurement.
- Check for skin parasites.
- Check the eyes and nose (any sign of discharge or inflammation).
- Check the cloaca (any signs of faecal staining).
- Check inside of the mouth (look for any worms, unusual spots, plaques, cheesy deposits etc).
- Check wings and legs, flexing the joints to look for breaks or asymmetry.
- Feel the keel. Is there good muscle cover?
- Pluck the ventral side of the bird, from the base of its neck to the under tail.
- Using the forceps pick up a little 'tent' of skin over the abdomen, and cut through the skin.
- Dissect the skin away from the abdomen and sternum. Cut the skin away all the way up the neck to the head.

➢ Assess abdominal fat birds in good condition will definitely had large quantities of solid fat obscuring the intestines. Birds in poor condition had little or no abdominal fat, so the intestines are easy to visualize as soon as open up the bird.

➢ Take a sample of any fat for toxicology before handle the bird further. Had the squares of Aluminium foil already torn and sitting ready (out of the way). Uses clean/brand new razor blade to cut this out. Use clean forceps and try to handle the sample as little as possible. Wrap the sample up without handling the inside of the foil, and make sure the package was small enough to drop into the shipper.

➢ Look for fluid, blood, clots, uric acid, any unusual spots, nodules, or growths etc. Take a sample for freezing.

➢ Cut through the pectoral muscles along either side of the ribs and sternum. Use the heavy scissors to cut the ribs on both sides. Use pliers to crack through the coracoids and furcula. Lift the sternum to expose the thoracic organs.

➢ Look for any inflamed/nasty looking wall to the intestine. Are there any red spots or little haemorrhages?

CHAPTER -05

RESULTS

Population count at Toawala and Dholewala study sites in 2001 to 2002

In 2001 breeding season a maximum population of Oriental White-backed Vulture at Toawala and Dholewala colony was counted as 1607 and 600 respectively. This count was taken in May when population including a few number of fledglings left (Table 1). While in second breeding season in 2002 from November 2001 to August 2002 a mean population of 1000 and 459 were observed at Toawala and Dholewala Colony respectively (Table 2). The both colonies were regularly observed on monthly basis. Total birds at Toawala including adults were 71.17%, sub adult were 20.90%, fledgling were 1.46%, while unknown aged groups were 1.46% (Table 32). Whereas at Dholewala Colony observed in 2002, comprised of adults were as 72.84%, sub adults were14.35%, fledglings and Juvenile were7.73%, while unknown aged groups were as 5.07% (Table 3).

Total area survey for the census was 39.50 Km at Toawala and 64.1 Km at Dholewala Colony. The density of vultures per Kilometer at Toawala was 40.68 vultures / Km in 2001 and 25.32 vultures / Km in 2002. However at Dholewala, density of vultures was 9.36 vultures / Km in 2001 and 7.16 Vultures / Km in 2002.

At Toawala Colony % decline in vultures/Kilometer area was 37.75% whereas at Dholewala, the decline was 23.50% (Table 8).

A sub sample of population estimation was taken in two intensive study transects at both sites in 2001 and 2002 on weekly basis. Total transect survey area per week was 7.67 Km and 5.1 Km at Toawala and Dholewala sites respectively. At Toawala Colony in 2001 a sum of 69.38% adult, 12.33% sub adults and fledglings/Juveniles as 18.27% were found (Table 4). On the other hand, at Dholewala Colony during 2001, 64.26% adults, 14.15% Sub adults and 21.57% fledglings/juveniles were found (Table 5). In 2002 at Toawala, adults were 37.79%, sub adults were 19.32%, fledglings/juveniles were 6.87% while unknown vultures were 0.09% (Table 6) .Whereas at Dholewala in 2002, adults were 66.07% sub adults were14.32%, fledglings/juveniles were 15.83% while unknown vultures were 3.60% (Table 7).

In conclusion 33.77% decline in vultures population was observed from 2001 to 2002 at Toawala. Whereas, 30.71% decline was observed during the same season at Dholewala.

Mean population and percentage declined was also observed in intensive study transects at Toawala and Dholewala Colony in 2001 and 2002. It was noted that from 2001 to 2002 decline was 65.17% and 38.21% at Toawala and Dholewala Colony respectively (Table 9).

For the sub sample in intensive study transects both at Toawala and Dholewala Colony, density of birds found at Toawala colony in 2001 was 70.53 vulture/Km. Of these 70.53 vultures per Km; adults were counted 48.89 vultures per Kilometer, sub adults were 8.74/Km and fledglings/juveniles were 12.91 vultures per Kilometer. While at Dholewala colony, Cumulative density was 43.13 vultures/ Km. Of these 27.84 vultures / Km were adults, sub adults were 6.07 vultures / Km and Fledgling and Juveniles were 9.21 vultures / Km.

While in 2002 season, cumulative density of vultures / Km noted at Toawala was 24.51 vultures/Km. A sum of 18.12 vultures / Km were in adult plumage, sub adults were 4.69 vultures / Km and fledglings/juveniles were 1.69 vultures/Km. While at Dholewala in 2002, cumulative density was noted as 26.66 vultures / Km. Of which adults were 17.64 vultures / Km, sub adults 3.72 vultures / Km and fledglings and juveniles were 4.31 vultures / Km (Table 10).

In conclusion during 2001 - 2002, both at Toawala and Dholewala, substantial decline in population of *Gyps bengalensis* was noted (Table 10).

Head drooping

A total of 18,194 Oriental White-backed Vultures were observed at two study sites (Toawala and Dholewala) in Punjab Province Pakistan for the purpose of head drooping analysis. Out of these 18,194 birds; a sum of 3,267 (17.95%) vultures displayed head drooping. Out of these vultures that displayed head drooping, adults were 1,990 (60.91%), sub adults were 376 (11.50%) and juvenile/fledglings were 900 (27.54%). Out of these drooping vultures, 1,582 (48.42%) birds displayed full drooping (as their head touching the claws in shade) while 1,685 (51.57%) displayed half drooping intensity. Out of 3,267 droopers, crop status was observed as full crop in 294 (8.99%), half full crop in 668 (20.44%), empty crop in 1,958 (59.93%) and unknown crop status was 254 (7.77%). Orientation to sun was also observed for drooping vultures and it was found as 227 (6.94%) facing to sun, 2,403 (73.55%) droopers backing to sun, 681 (20.84%) were lateral to sun direction, 56 (1.71%) were those in which sun was observed above the head of the droopers. Perched substrate was also observed for each of drooping bird and it was found that drooping birds found on tree were 3,055 (93.51%) and droopers on ground were 212 (6.48%). Mean temperature was also recorded for droopers and it was found 31.11 Celsius and 33.19 Celsius at Toawala and Dholewala colony in 2001 and 35.40 Celsius and 34.02 Celsius at Toawala and Dholewala study sites respectively in 2002 (Table 11).

Head drooping in first season 2001

Table 12 and 13 show that in 2001 season, a total of 5,820 and 2,839 oriented White-backed Vultures were observed at Toawala and Dholewala respectively. Of which 1,131 (19.434%) and

486 (17.12%) vultures displayed head drooping at Toawala and Dholewala respectively. From these, age class of each drooping vulture was identified and it was found that 670 (59.23%) and 233 (47.94%) vultures were adult, sub adult were 104 (9.19%) and 30 (6.17%) and juvenile / fledgling were 357 (31.56%) and 223 (45.88%) at Toawala and Dholewala respectively. Drooping intensity was also recorded and full drooping vultures were 271 (23.96%) and 422 (86.83%), while half droopers were 860 (76.03%) and 64 (13.16%) at Toawala and Dholewala respectively. Crop status was also noted, as full crop were 125 (11.05%) & 13 (2.67%) vultures, half filled crop were 288 (25.46%) and 94 (19.34%), empty crop vultures were exceedingly higher 713 (63.04%) and 293 (60.28%) while unknown were 5 (0.04%) and 86 (17.69%) at Toawala and Dholewala respectively.

Orientation of the birds were also taken into account and it was observed that vultures facing to sun were 31 (2.74%) and 68 (13.99%) while backing to sun were exceedingly higher 855 (75.59%) and 316 (65.02%), lateral to sun direction were189 (16.71%) and 10 (20.98%), while sun above to vultures were observed in 56 (4.95%) and 0 (0%) at Toawala and Dholewala study sites respectively. Parching substrate for all vultures displayed drooping was also taken into account and it was observed in 1,112 (98.32%) and 395 (81.27%) vultures on tree, while rest 19 (1.67%) and 91 (18.72%) were on ground in Toawala and Dholewala sites respectively (Table 12, Table 13).

Head drooping in second season 2002

In second season 2002, Table 14 and 15 shows a total of 5,146 and 4,389 vultures were observed. Out of which 824 (16.01%) and 826 (18.81%) vultures displayed head drooping at Toawala and Dholewala Colony. Age class of drooping vultures was also observed and it was found that 479 (58.13%) and 608 (73.60%) vultures were adult, sub adult were 193 (23.42%) and 49 (5.93%) while juvenile and fledglings were 152 (18.44%) and 168 (20.33%) at Toawala and Dholewala study sites respectively. Drooping intensity was observed and full drooping vultures were found 175 (21.23%) and 714 (86.44%) and half droopers were 649 (78.76%) and 112 (13.55%) at Toawala and Dholewala respectively. Crop status was also observed and it was found that vultures with full crop were 133 (16.14%) and 23 (2.78%), half filled crop were 197 (23.90%) and 89 (10.77%) and empty crop vultures were exceedingly higher 491 (59.58%) and 461 (55.81%). Crop status of some vultures were not observed as 0 (0%) and 153 (18.52%) at Toawala and Dholewala colony respectively.

Orientation to the sun was also recorded for each vulture and it was found that vultures facing to sun were 53 (6.43%) and 75 (9.07%), vultures backing to sun were exceedingly higher as 526 (63.83%) and 606 (73.36%) and lateral to sun direction were 245 (29.73%) and 145 (17.55%) at Toawala and Dholewala respectively. Perched substrate was also observed as 763 (92.59%) and

785 (95.03%) vultures were on trees while on ground it was observed in 61 (7.40%) and 41 (4.96%) at Toawala and Dholewala study sites respectively (Table14, Table 15).

Correlation of droopers with different parameters at Toawala in 2001

In present study, age class (adult, sub adult, fledgling/ juvenile), drooping intensity (full droop intensity, half droop intensity), crop status (full crop, half filled crop, empty crop), orientation to the sun (front, back, lateral, above), the perched substrate (tree, ground) and temperature were considered to be independent variables parameters. The effect of these variables was observed on the droopers (dependent variable). The results regarding the correlation between different parameters are shown in (Table 16).

The data (n = 24) showed highly significant positive correlation at (p<0.001) between the droopers and adult (r = 0.993), droopers with sub adult (r = 0.573), with juvenile (r = 0.965), with full droop (r = 0.910), with half droop (r = 0.994), with full crop (r = 0.839), with half filled crop (r = 0.981), with empty crop (r = 0.987), with backing to the sun (r = 0.976), with lateral (r = 0.834), with perched substrate as on tree (r = 1.00) and droopers with temperature (r = 0.885, p<0.001) was highly significant positive correlation.

While droopers showed fairly significant positive correlation with the sub adult (r = 0.573, p<0.01) and a significant correlation with perched substrate as ground (r = 0.407).

Droopers showed non significant correlation among the unknown crop (r = 0.125, p>0.05) and with orientation as facing to sun and for unknown crop status (r = 0.239, r = 0.239, r = 0.125, p>0.05) respectively (Table 16).

Correlation of droopers with different parameters at Dholewala in 2001

The results regarding the correlation of the droopers with the other independent variables at Dholewala Colony (n = 19) are shown in Table 17. Data (n = 19) showed that there was a highly significant positive correlation at (p<0.001) between the droopers and adult (r = 0.951), droopers and sub adult, droopers and juvenile (p<0.001, r = 0.876, r = 0.980) respectively. Between droopers and drooping intensity as full droop was (p<0.001, r = 0.994). Droopers with crop status as half filled crop, empty crop and unknown crop was (p<0.001, r = 0.919, r = 0.980, r = 0.934) respectively. Droopers with orientation as facing to sun, backing to sun and lateral to sun was (p<0.001, r=0.935, r = 0.989, r = 0.979) respectively. Droopers with perched substrate as with tree and with ground (p<0.001, r = 0.989, r = 0.834) respectively (Table17).

There was a non significant correlation between droopers and half droop intensity (P>0.05, r = 0.046) and with full crop (P>0.05, r = 0.301).

There was a highly significant positive correlation between the droopers and the temperature (P<0.001, r = 0.859) observed.

Correlation of droopers with different parameters at Toawala in 2002

The results regarding to the correlation between droopers and different parameters (independent variables) at Toawala study site in 2002 were shown in Table 18.

The data (n = 34) showed that there was a highly significant positive correlation between the droopers and adult (p<0.001, r = 0.995), droopers and sub adults (p<0.001, r = 0.956) and Droopers with Juvenile (p<0.001, r = 0.919). Drooping intensity also showed a highly significant positive correlation of droopers and full drooping intensity (p<0.001, r = 0.962) and half drooping intensity (p<0.001 r = 0.995). The correlation between the crop status and the droopers showed highly significant correlation with full crop, half filled crop and empty crop (p<0.001, r = 0.678, r = 0.988, r = 0.974) respectively.

It is evident that correlation of droopers with orientation as backing to sun and lateral (p<0.001, r = 0.980, r = 0.859) were highly significant. The association of droopers with the perched substrate is highly significantly correlated with tree and ground (p<0.001, r = 0.998, r = 0.749) respectively. It is clear from the data that head drooping is highly positively correlated with the temperature (n = 34, p<0.001, r = 0.870).

There is a fairly significant positive correlation between the droopers and orientation facing to sun (p<0.001, r = 0.486) Table 18.

Correlation of droopers with different parameters at Dholewala in 2002

The data (n = 33) showed that correlation between the droopers (dependent variable) and the independent variables at Dholewala in 2002 in the Table 19.

The data (n = 33) showed that droopers showed a highly significant positive correlation with age class as adult, sub adult and juvenile (p<0.001, r = 0.988, r = 0.845, r = 0.581) respectively. It is clear that full drooping intensity is highly significantly correlated with droopers (p<0.001, r = 0.988).Droopers showed a highly significant positive correlation with full crop, empty crop and unknown crop (p<0.001, r = 0.591, r = 0.808, r = 0.589) respectively. It is evident that orientation to the sun is highly significant positive correlation as facing to sun (r = 0.822), backing to sun (r = 0.987), lateral direction to sun (r = 0.744) at (p<0.001).There was a highly significant positive correlation in perched substrate on tree (r = 0.991). Temperature was highly significant positively correlated with the droopers (p<0.001, r = 0.809) Table 19.

There is a negative correlation of droopers with the perched substrate on ground (r = -0.333) Table 19.

Crop status of non-droopers at Toawala in 2001

A total of 6,283 vultures were observed at Toawala in 2001 for the drooping data of which 5,152 (81.99%) vultures display no head drooping posture. Of these vultures 656 (12.73%) had full crop while half filled crop was accounted in1,068 (20.72%) vulture. The percentage of empty crop was highest as 3,349 (65%) of Oriental White-backed Vultures studied. While crop status for only 79 (1.53%) vultures was not seen (Table 20).

Crop status on non droopers at Dholewala in 2001

In Dholewala study site, a total of 3,678 Oriental White-backed Vultures were observed for drooping data. Out of which 3,192 (86.78%) vultures do not displays head drooping posture. Crop status of these vultures was taken into account and this showed that 411 (12.87%) had full crop and 961 (30.11%) had half filled crop. Empty Crop was observed in 1465 (45.90%) vultures, which was a high number. Crop status for 355 (11.12%) vultures was not possible to observed (Table 21).

Crop status of non droopers at Toawala in 2002

In 2002, at Toawala study site, a total of 5,603 Oriental White-backed Vultures were taken observed for the head drooping data. Out of these 4,745 (85%) vultures do not displays head drooping posture. Crop status of each vulture was observed and it was noted that 991 (20.88%) vulture had full crop, while 1,014 (21.36%) vultures had half filled crop. High number of birds was observed with empty crop 2,724 (57.40%). Crop status was not observed in 16 (0.34%) of oriental White-backed Vultures (Table 22).

Crop status of non-droopers at Dholewala in 2002

A total of 5,338 Oriented White-backed Vultures were observed. Out of these, 4,293 (80.42%) vultures do not displays head drooping posture. Out of these non droopers, crop status was also observed and it was noted that 307 (7.15%) vultures were with full crop and 617 (14.37%) vultures were with half filled crop. The percentage of empty crop was higher as in 2,056 (47.89%). While crop status of 1,313 (30.58%) vultures was unknown (Table 23).

Correlation of non droopers with crop status at Toawala in 2001

The data (n = 33) indicate that there was a high highly significant positive correlation between the non droopers and the crop status as with full crop (r = 0.770, p<0.01), with non droopers and with half filled crop (r = 0.861, P<0.001), with empty crop (r = 0.964, p<0.001).

While there was a negative correlation between the non droopers and unknown crop status (r = -0.308 P>0.05) Table 24.

Correlation of non droopers with crop status at Dholewala in 2001

Data (n = 30) showed that there was a highly significant positive correlation between the non droopers and crop status as half filled crop (r = 0.705, p<0.001).While for empty crop highly significant positive correlation was observed with non droopers (r = 0.856, p<0.001) and for unknown crop status with non droopers, there was a high significant positive correlation (r = 0.758, p<0.001). There was a significant positive correlation between the non-droopers and full crop (r = 0.333, p<0.05) Table 25.

Correlation of non droopers with crop status at Toawala in 2002

In 2002 at Toawala, data (n = 36) showed that there was a highly significant positive correlation between the non droopers and full crop (r = 0.906, p<0.001), half filled crop (r = 0.980) and with empty crop (r = 0.976, p<0.001).

While there was a negative correlation among non-droopers and vultures with unknown crop status (r = -0.308, p>0.05) same as observed at Toawala in 2001 (Table 26).

Correlation of non droopers with crop status at Dholewala in 2002

The data (n = 30) showed, that there was a highly significant positive correlation between the non droopers with crop status as half filled crop (r = 0.705, p<0.001), with empty crop (r = 0.850, p<0.001) and with unknown crop status (r = 0.758, p<0.001) same as observed in 2001 at Dholewala.

While there was a significant correlation between the full crop status and non droopers (r = 0.333, p<0.05) Table 27.

Mortality analysis

A total of 869 dead, sick and injured Oriental White-backed Vultures were found from two study sites Toawala and Dholewala during the study period from December 2000 to August 2002. In relation to that a total of 389 (44.76%) were adult, 116 (13.34%) were sub adult, 353 (40.62%) were nestling/fledgling and 12 (1.38%) were Unknown or unidentified (which were found as only bones, feather or wings piles).Adult and sub adult mortality rate was consistently high during the study period while nestling mortality was low until April and May in both breeding seasons. In these months mortality of nestling increased due to fledgling fatalities. This is believed to be attributed to naivete (e.g. unable to take off after their first flight). The proportion of each age class known to have died within the study sites during the study period is summarized in (Table32).

Mortality at Toawala in December 2000 - July 2001

A total of 141 dead Oriental White-backed Vultures are collected and removed from the study site at Toawala in first season from December 2000-July 2001. During this breeding season, a sum of 141 dead, adults were accounted 35 (24.82%), sub adult 7 (4.96%) while nestling/fledgling mortality was exceedingly high as 97 (68.79%) and for Unidentified vultures as 2(1.41%). The fledgling mortality rate was high immediately after fledgling. The proportion of the age class of Oriental White-backed Vultures is summarized in Table 28.

Mortality at Dholewala in November 2000 - July 2001

From the second study site in Punjab Province of Pakistan, a total of 223 dead Oriental White-backed Vultures were collected in first breeding season during the study period from November 2000-July 2001. From these, a total of 78 (34.97%) were adult, 18 (8.07%) were sub adult, 123 (55.15%) were Nestling/fledgling and 4 (1.79%) were unidentified vultures were collected and removed from the study site. Also fledgling mortality was exceedingly high similar to that of Toawala. The percentage and proportion of age class is given in Table 29.

Mortality at Toawala in October 2001 - August 2002

In the second breeding season, a sum of 185 dead, sick and injured Oriental White-backed Vultures were found at Toawala. From these, adult were 78 (42.16%), sub-adult were 45 (24.32%), Nestling/fledglings were 57 (30.81%) and Unidentified were 6 (3.24%). The proportion of the age class of Oriental White-backed Vultures is summarized in Table 30.

Mortality at Dholewala in October 2001 - August 2002

From the second study site in Punjab Province of Pakistan, a sum of 320 dead, sick and injured Oriental White-backed Vultures were found and removed in second breeding season during October 2000- August 2001.From these, 198 (61.87%) were adult, sub adults were 46 (14.37%) and Nestling/fledglings were 76 (23.75). The percentage and proportion of age class is given in Table 31.

Mortality rate

Adult annual mortality rate was calculated 7.7% at Toawala colony in first breeding season while 9.9% at Dholewala during December 2000 to July 2001. While adult and sub adult cumulative mortality rate calculated at Toawala was 9.3% and at Dholewala is 12.2% (Table 33).

In second breeding season during the study period from October 2001 to 2002, annual adult mortality rate was calculated as 10.6% at Toawala, while at Dholewala it was 26.2%. The adult and sub adult combine annual mortality rate was calculated as 16.7% and 32.3% at Toawala and Dholewala respectively (Table 33).

Gout and omental fat reserves

An assessment of gout was made in 266 freshly dead Oriental White-backed Vultures through necropsies. Of these 266 vultures, adults were 102 (38.34%), sub adults were 53 (19.92%), fledglings and juvenile were 83 (31.20%) and nestlings were 25 (9.39%) had gout (Table 40). All these necropsies were done from December 2000-July 2001 and October 2001 to August 2002 in two breeding seasons. Of these 57 and 76 necropsies were done at Toawala in first and second breeding season respectively. While 55 and 78 necropsies were done at Dholewala in first and second breeding season respectively. Out of these 266 necropsies, a total of 158 (59.39%) dead vultures had the signs of visceral gout (Table 38). Omental fat reserves were found intact in 200 (75.18%) of the dead Oriental White-backed Vultures (Table 38). All other dead vultures without showing gout were died from dehydration, heat stress, injury and other causes presumed to relate to the naivete of the birds during the post fledgling period.

Gout and omental fat reserves at Toawala in 2001

At Toawala colony, in first season a total of 57 dead vultures were resected for the assessment of gout. From these only 6 (10.53%) showed the signs of visceral gout (Table 38). Gout was found in 1 (25%) of the adults, sub adults were 1 (25%), fledgling were 4 (10.31%) and in nestlings (n = 8) gout was not observed (Table 34). An assessment of sex was made in 7 of these 57 dead vultures, out of these male were not observed having gout while 1 (14.28%) female was observed with gout (Table 39). Age class of these 57 dead vultures is shown in Table 40. Omental fat reserves were found intact in 51 (89.47%) vultures (Table 38).

Gout and omental fat reserves at Dholewala in 2001

At Dholewala colony, in first breeding season, assessment of gout was made on 55 dead vultures of these 31 (56.36) had the signs of visceral gout (Table 38). Gout was found in 16 (88.9%) adults, sub adults were 2 (50.0%), fledglings were 9 (47.4%) and 3 (23.1%) were nestling (Table 35). An assessment of sex was made in 12 of these 55 dead vultures. Out of these 12 vultures, male had 6 (50.0%) visceral gout while in female 2 (16.66%) gout was observed (Table39). Age class of these 55 dead vultures is shown in Table 40. Omental fat reserves were found intact in 37 (67.27%) of the dead vultures (Table 38).

Gout and omental fat reserves at Toawala in 2002

At Toawala colony, in second breeding season, a total of 76 dead vultures were opened for the assessment of the visceral gout and omental fat reserves. Out of these, gout was found in 56 (73.68%) of the dead Oriental White-backed Vultures resected for assessment of visceral gout (Table 38). From these 56 dead birds with gout, 32 (86.5%) were adults, sub adults were 17 (77.3%), fledglings were 8 (47.1%) and no nestling was observed with gout (Table 36). An assessment of sex was made in 47 vultures. From these 32 (68.10%) were male and 11 (23.40%) were female had signs of visceral gout (Table 39). Age class of these 76 dead vultures is shown in Table 40. An assessment of fat reserves was made in 76 vultures, out of these 52 (68.42%) had the omental fat reserves intact in the body (Table 38).

Gout and omental fat reserves at Dholewala in 2002

In second breeding season at Dholewala, Assessment of the visceral gout and omental fat reserves was made in 78 dead vultures through necropsy. Gout was found in 65 (83.33%) of these dead Oriental White-backed Vultures resected (Table 38). Out of these 65 dead vultures with gout, 38 (88.4%) were adults, sub adults were 21 (91.3%), fledglings were 2 (25.0%) and nestling were 4 (100%) with signs of visceral gout (Table 37). An assessment of sex was made in 50 vultures. From these 28 (56.0%) were male and 17 (34.0%) were female had the signs of visceral gout (Table 39). Age class of these 78 dead vultures is shown in Table 40. Assessment for the omental fat reserves was found in 60 (76.92%) of the dead vultures opened for necropsy (Table 38).

38

Table 1. Toawala And Dholewala (Whole Site) Count in 2001

Month	Total	Adult	Sub-adult	F/J	Unknown
Toawala	1607	0	0	0	0
Dholewala	600	0	0	0	0

Table 2. Toawala (Whole Site) Count in 2001-2002

Month	Total	Adult	Sub-adult	F/J	Unknown
November	1232	885	231	0	116
December	1068	1003	65	0	0
January	1205	979	226	0	0
February	1231	1006	225	0	0
March	1135	840	245	50	0
April	1052	676	247	129	0
May	711	367	210	134	0
June	884	438	277	139	30
July	806	442	239	125	0
August	674	480	125	69	0
Total	9998	7116	2090	646	146
Mean	999.80	711.60	209.00	64.60	14.60
% age	100	71.17	20.9	6.46	1.46

Table 3. Dholewala (Whole Site) Count in 2001-2002

Month	Total	Adult	Sub-adult
November	305	215	90
December	732	613	119
January	552	460	92
February	612	517	95
March	626	462	78
April	725	332	125
May	281	173	29
June	214	151	6
July	224	157	15
August	321	265	10
Total	4592	3345	659
Mean	459.20	334.50	65.90
% age	100	72.84	14.35

Table 4. Weekly Count at Toawala study site in 2001

Month	Total	Adult	Sub Adult	Flg/Juv
May	375.5	243.5	58	74
June	640.33	465	80	95.33
July	605.6	416.6	62	127
Total	1621.43	1125.1	200	296.33
Mean	540.48	375.03	66.67	98.78
%age	100	69.38	12.33	18.27

Table 5. Weekly Count at Dholewala study site in 2001

Month	Total	Adult	Sub Adult	Flg/Juv
June	79	36	6	37
July	295	180	52	63
August	286.5	208.5	35.5	42.5
Total	660.5	424.5	93.5	142.5
Mean	220.17	141.50	31.16	47.50
%age	100	64.26	14.15	21.57

Table 6. Weekly Count at Toawala study site in 2002

Month	Total	Adult	Sub Adult	Flg/Juv	Unknown
November	180.5	151.5	23.5	5.5	0
December	177.5	160.5	16.5	2	0
January	215	185.5	29.5	0	0
February	219	188	31	0	0
March	244	188	55	1	0
April	200.5	136.5	54	10	0
May	163.11	88.7	44.25	30	0
June	171.61	93.87	48.62	27.5	1.62
July	155.6	97.62	26.98	31	0
August	150.4	95	33.4	22	0
Total	1877.22	1385.19	362.75	129	1.62
Mean	187.72	138.52	36.28	12.90	0.16
%age	100	73.79	19.32	6.87	0.09

Table 7. Weekly Count at Dholewala study site in 2002

Month	Total	Adult	Sub Adult	Flg/Juv	Unknown
October	256	153	61.5	41.5	0
November	163	63	11	89	0
December	160.75	132.75	28	0	0
January	145.75	111.5	34.25	0	0
February	147.5	121.5	26	0	0
March	168.75	107	19.5	0	42.25
April	141.25	71.75	17.5	52	0
May	82.5	45.5	9	25.5	0
June	65.5	54	3	8.5	0
July	97.75	84	2.75	11	0
August	65	43	1.5	9	11.5
Total	1493.75	987	214	236.5	53.75
Mean	135.80	89.73	19.45	21.50	4.89
%age	100	66.07	14.32	15.83	3.6

Table 8. Decline of OWBV/ km at Toawala and Dholewala Whole sites from 2001-2002

Sites	Mean Total Vulture	Total area	Vulture/Km.	% Decline Vulture/Km.2001-2002
Toawala 2001	1600	39.5	40.68	37.75
Toawala 2002	1000	39.5	25.32	
Dholewala 2001	600	64.1	9.32	23.5
Dholewala 2002	459	64.1	7.16	

Table 9. Decline of OWBV/ km. at Toawala and Dholewala Whole sites from 2001-2002

Sites	mean vultures at TW2001	mean vultures at DW2001	mean vultures at TW2002	mean vultures at DW2002	% Decline at TW2001-2002	% Decline at DW2001-2002
Whole sites	1607	600	1000	459	33.77	30.71
Weekly sites	540	220	188	136	65.17	38.21

Table 10. Decline in age class of OWBV/ km at Toawala and Dholewala sites in intensive transects from 2001-2002.

Sites	Mean Total	Mean Adult	Mean S-adult	Mean Fledgling/Juvenile
Toawala 2001	541	375	67	99
Toawala 2002	188	139	36	13
Total Area	7.67	7.67	7.67	7.67
Vulture/Km.at Toawala 2001	70.53	48.89	8.74	12.91
Vulture/Km.at Toawala 2002	24.51	18.12	4.69	1.69
% Decline Vulture/Km.2001-2002	65.18	62.93	46.33	86.90
Dholewala 2001	220	142	31	47
Dholewala 2002	136	90	19	22
Total Area	5.1	5.1	5.1	5.1
Vulture/Km.at Dholewala 2001	43.13	27.84	6.07	9.21
Vulture/Km.at Dholewala 2001	26.66	17.64	3.72	4.31
% Decline Vulture/Km.2001-2002	38.18	36.63	38.71	53.20

Table 11. Comparison of Head Drooping Analysis at Toawala and Dholewala study sites in 2001-2002

SITE	NO. OF Obs.	NO. OF Droopers	Age Class Adult No.	Age Class Sub-Adult No.	Age Class Juvenile No.	Drooping Intensity Full No.	Drooping Intensity Half No.	Crop Status Full No.	Crop Status Half No.	Crop Status Empty No.	Crop Status Unknown No.	Orientation to Sun Front No.	Orientation to Sun Back No.	Orientation to Sun Lateral No.	Orientation to Sun Above No.	Orientation to Sun Unknown No.	Perch Substrat Tree No.	Perch Substrat Ground No.	Mean Temprature
TOAWALA 01	5820	1131	670	104	357	271	860	125	288	713	5	31	855	189	56	0	1112	19	31.105
TOAWALA 02	5146	824	479	193	152	175	649	133	197	491	0	53	526	245	0	0	763	61	35.39
TOTAL	10966	1955	1149	297	509	446	1509	258	485	1204	5	84	1381	434	56	0	1875	80	33.25
%AGE	100	17.82	58.77	15.19	26.04	22.81	77.18	13.2	24.81	61.58	0.25	4.29	75.75	22.19	2.86	0	95.9	4.09	33.25
DHOLEWALA 01	2839	486	233	30	223	422	64	13	94	293	86	68	316	102	0	0	395	91	33.19
DHOLEWALA 02	4389	826	608	49	168	714	112	23	89	461	153	75	606	145	0	0	785	41	34.02
TOTAL	7228	1312	841	79	391	1136	176	36	183	754	239	143	922	247	0	0	1180	132	33.6
%AGE	100	18.15	64.1	6.02	29.8	86.58	13.41	2.74	13.94	57.46	18.97	10.89	70.27	18.82	0	0	89.93	10.06	33.6
GRAND TOTAL	18194	3267	1990	376	900	1582	1685	294	668	1958	254	227	2403	681	56	0	3055	212	33.42
%AGE	100	17.95	60.91	11.5	27.54	48.42	51.57	8.99	20.44	59.93	7.77	6.94	73.55	20.84	1.71	0	93.51	6.48	33.42

Table 12. Head Drooping Analysis at Toawala study sites in 2001.

MONTH	NO OF Obs	NO OF Droopers	Age Class: Adult No	Adult %age	S-Adult No	S-Adult %age	Juvenile No	Juvenile %age	Intensity Full No	Full %age	Intensity Half No	Half %age	Crop Full No	Crop Full %age	Crop Half No	Crop Half %age	Crop Empty No	Crop Empty %age	Crop Unknown No	Crop Unknown %age	Front No	Front %age	Back No	Back %age	Lateral No	Lateral %age	Above No	Above %age	Orient Unknown No	Orient Unknown %age	Tree No	Tree %age	Ground No	Ground %age	%age Droopers	Mean Temp
FEB	897	6	6	100.00	0	0.00	0	0.00	1	16.70	5	83.30	0	0.00	1	16.70	5	83.30	0	0.00	6	100.00	0	0.00	0	0.00	0	0.00	0	0.00	6	100.00	0	0.00	0.68	21.21
MAR	432	6	6	100.00	0	0.00	0	0.00	2	33.30	4	66.70	0	0.00	0	0.00	6	100.00	0	0.00	2	33.30	3	50.00	1	16.70	0	0.00	0	0.00	6	100.00	0	0.00	1.38	25.57
APR	918	232	134	57.75	52	22.40	46	19.82	83	35.80	149	64.20	44	18.96	70	30.20	113	48.70	5	2.15	7	3.00	168	72.41	57	24.60	0	0.00	0	0.00	232	100.00	0	0.00	25.27	33.78
MAY	1397	315	167	53.00	28	8.90	120	38.10	70	72.20	245	77.80	15	48.00	81	69.51	219	69.50	0	0.00	13	4.12	193	61.30	73	23.20	36	11.40	0	0.00	315	100.00	0	0.00	22.56	34.39
JUN	1584	431	274	63.60	18	4.20	139	32.20	78	18.10	353	81.90	59	13.70	95	22.40	277	64.30	0	0.00	2	0.50	382	88.60	47	10.90	0	0.00	0	0.00	422	97.91	9	2.10	27.20	37.42
JUL	592	141	83	58.86	6	4.25	52	36.87	37	26.24	104	73.75	7	4.96	41	29.07	93	65.95	0	0.00	1	0.70	109	77.30	11	7.80	20	14.18	0	0.00	131	92.90	10	7.09	23.81	34.26
TOTAL	**5820**	**1131**	**670**	**59.23**	**104**	**9.19**	**357**	**31.56**	**271**	**23.96**	**860**	**76.03**	**125**	**11.05**	**288**	**25.46**	**713**	**63.04**	**5**	**0.04**	**31**	**2.74**	**855**	**75.59**	**189**	**16.71**	**56**	**4.95**	**0**	**0.00**	**1112**	**98.32**	**19**	**1.67**	**19.43**	**31.11**

Table 13. Head Drooping Analysis at Dholewala study sites in 2001

MONTH	NO. OF Obs	NO. OF Droopers	%age of Age Class of Droopers						%age of Drooping Intensity				Percentage of Crop Status								Orientation to Sun										%age of Perch Substrat				%age Droopers	Mean Temp
			Adult		S-Adult		Juvenile		Full		Half		Full		Half		Empty		Unknown		Front		Back		Lateral		Above		Unknown		Tree		Ground			
			No.	%age	No.	%age	No.	%age	No.	%age	No.	%age	No.	%age	No.	%age	No.	%age	No.	%age	No.	%age	No.	%age	No.	%age	No.	%age	No.	%age	No.	% age	No.	% age		
MAR	448	4	3	75.00	0	0.00	1	25.00	1	25.00	3	4.00	1	25.00	1	25.00	0	0.00	2	50.00	2	50.00	1	25.00	1	25.00	0	0.00	0	0.00	4	100.00	0	0.00	0.89	25.58
APR	527	49	35	71.43	6	12.24	8	16.33	38	77.55	11	15.40	2	4.08	10	20.41	33	67.35	4	8.16	9	18.37	28	57.14	12	24.49	0	0.00	0	0.00	45	91.84	4	8.16	9.30	35.06
MAY	737	201	74	36.82	10	4.98	117	58.21	194	96.52	7	19.01	0	0.00	30	14.93	142	70.65	29	14.43	24	11.94	134	66.67	43	21.39	0	0.00	0	0.00	176	87.56	25	12.44	27.27	39.26
JUN	818	169	76	44.97	13	7.69	80	47.34	153	90.53	16	35.58	8	4.73	31	18.34	94	55.62	36	21.30	29	17.16	100	59.17	40	23.67	0	0.00	0	0.00	122	72.19	47	27.81	20.66	35.76
JULY	309	63	45	71.43	1	1.59	17	26.98	36	57.14	27	37.80	2	3.17	22	34.92	24	38.10	15	23.81	4	6.35	53	84.13	6	9.52	0	0.00	0	0.00	48	76.19	15	23.81	20.39	30.31
TOTAL	2839	486	233	47.94	30	6.17	223	45.88	422	86.83	64	13.16	13	2.67	94	19.34	293	60.28	86	17.69	68	13.99	316	65.02	102	20.98	0	0.00	0	0.00	395	81.27	91	18.72	17.12	33.19

Table 14. Head Drooping Analysis at Toawala study sites in 2002

MONTH	NO. OF Obs	NO. OF Droopers	%age of Age Class of Droopers						%age of Drooping Intensity				Percentage of Crop Status								Orientation to Sun										%age of Perch Substrat				%age Droopers	Mean Temp
			Adult		S-Adult		Juvenile		Full		Half		Full		Half		Empty		Unknown		Front		Back		Lateral		Above		Unknown		Tree		Ground			
			No.	%age	No.	%age	No.	%age	No.	%age	No.	%age	No.	%age	No.	%age	No.	%age	No.	%age	No.	%age	No.	%age	No.	%age	No.	%age	No.	%age	No.	%age	No.	%age		
MAR	234	8	6	75.00	2	25.00	0	0.00	1	12.50	7	87.50	2	25.00	1	12.50	5	62.50	0	0.00	2	25.00	3	37.50	3	37.50	0	0.00	0	0.00	8	100.00	0	0.00	3.41	31.03
APR	864	78	47	60.25	21	26.92	10	12.82	18	23.07	60	76.92	10	12.82	21	26.92	47	60.25	0	0.00	12	15.38	34	43.58	32	41.02	0	0.00	0	0.00	73	93.58	5	6.41	9.02	34.65
MAY	844	153	88	57.51	23	15.03	42	27.45	26	16.99	127	83.00	17	11.11	34	22.22	101	66.01	0	0.00	5	3.26	97	63.39	51	33.33	0	0.00	0	0.00	132	86.27	21	13.72	18.12	37.81
JUNE	1134	276	167	60.50	65	23.55	44	15.94	74	26.81	202	73.18	26	9.42	62	22.46	186	67.39	0	0.00	10	3.62	211	76.44	55	19.93	0	0.00	0	0.00	262	94.92	14	5.07	24.33	38.43
JULY	1170	202	107	52.97	55	27.22	40	19.80	41	20.29	161	79.70	55	27.22	47	23.26	100	49.50	0	0.00	13	6.43	115	56.93	74	36.63	0	0.00	0	0.00	185	91.58	17	8.41	17.26	35.50
AUG	900	107	64	59.81	27	25.23	16	14.95	15	14.01	92	85.98	23	21.49	32	29.90	52	48.59	0	0.00	11	10.28	66	61.68	30	28.03	0	0.00	0	0.00	103	96.26	4	3.73	11.88	34.97
TOTAL	5146	824	479	58.13	193	23.42	152	18.44	175	21.23	649	78.76	133	16.14	197	23.90	491	59.58	0	0.00	53	6.43	526	63.83	245	29.73	0	0.00	0	0.00	763	92.59	61	7.40	15.31	35.40

47

Table 15. Head Drooping Analysis at Dholewala study sites in 2002

MONTH	NO. OF Obs	NO. OF Droopers	Adult No.	Adult %age	S-Adult No.	S-Adult %age	Juvenile No.	Juvenile %age	Full No.	Full %age	Half No.	Half %age	Full No.	Full %age	Half No.	Half %age	Empty No.	Empty %age	Unknown No.	Unknown %age	Front No.	Front %age	Back No.	Back %age	Lateral No.	Lateral %age	Above No.	Above %age	Unknown No.	Unknown %age	Tree No.	Tree %age	Ground No.	Ground %age	%age Droopers	Mean Temp
			%age of Age Class of Droopers						%age of Drooping Intensity				Percentage of Crop Status								Orientation to Sun										%age of Perch Substrat					
MAR	134	28	19	67.85	6	21.42	2	7.14	22	78.57	6	21.42	0	0.00	3	10.71	20	71.42	5	17.85	1	3.57	23	82.21	4	14.28	0	0.00	0	0.00	28	100.0	0	0.00	22.58	31.03
APR	553	133	86	64.66	9	6.76	38	28.57	103	77.44	30	22.55	4	3.00	14	10.52	85	63.90	30	22.55	15	11.27	97	72.93	21	15.78	0	0.00	0	0.00	131	98.5	2	1.50	24.05	35.06
MAY	423	69	31	44.92	0	0.00	38	55.70	39	56.52	30	43.47	4	5.79	14	20.28	15	21.73	36	52.17	3	4.34	38	55.07	28	40.57	0	0.00	0	0.00	37	53.6	32	46.38	17.40	39.26
JUN	1520	261	202	77.39	15	5.74	44	16.85	231	88.51	30	11.49	4	1.53	8	3.06	102	39.08	47	18.01	43	16.47	184	32.18	34	13.02	0	0.00	0	0.00	258	98.9	3	1.15	17.10	35.76
JUL	829	172	144	83.72	10	5.81	18	10.46	167	97.09	5	2.91	7	4.06	7	4.06	136	79.06	22	12.79	6	3.48	142	82.55	24	13.95	0	0.00	0	0.00	168	97.7	4	2.32	20.74	29.86
AUG	934	163	126	77.30	9	5.52	28	17.17	152	93.25	11	6.74	4	2.45	43	26.38	103	65.50	13	7.97	7	4.29	122	74.84	34	20.85	0	0.00	0	0.00	163	100.0	0	0.00	17.45	33.15
TOTAL	4389	826	608	73.60	49	5.93	168	20.33	714	86.44	112	13.55	23	2.78	89	10.77	461	55.81	153	18.52	75	9.07	606	73.36	145	17.55	0	0.00	0	0.00	785	95.0	41	4.96	18.81	34.02

48

Table 16. Correlation of head droopers with other parameters at Toawala in 2001

	Droopers	Adult	Sub Adult	Juvenile	Full Droop	Half Droop	Full Crop	Half Crop	Empty Crop	Unknown	Front	Back	Lateral	Above	Tree	Ground
Adult	0.993***															
Sub Adult	0.573**	0.524**														
Juvenile	0.965***	0.949***	0.400*													
Full Droop	0.910***	0.882***	0.847***	0.805***												
Half Droop	0.994***	0.993***	0.481**	0.979***	0.858***											
Full Crop	0.839***	0.874***	0.655***	0.678***	0.842***	0.812***										
Half Crop	0.981***	0.959***	0.695***	0.927***	0.968***	0.955***	0.825***									
Empty Crop	0.987***	0.977***	0.463*	0.994***	0.847***	0.995***	0.755***	0.952***								
Unknown	0.125	0.106	0.840***	-0.112	0.495*	0.020	0.457*	0.264	-0.026							
Front	0.239	0.140	0.507*	0.276	0.346	0.203	-0.032	0.315	0.256	0.198						
Back	0.976***	0.995***	0.476*	0.926***	0.850***	0.980***	0.891***	0.932***	0.958***	0.088	0.045					
Lateral	0.834***	0.772***	0.830***	0.782***	0.913***	0.787***	0.643**	0.894***	0.803***	0.397*	0.670***	0.709***				
Above	0.276	0.180	0.102	0.450*	0.255	0.273	-0.256	0.335	0.364*	-0.298	0.572**	0.105	0.450*			
Tree	1.000***	0.991***	0.586**	0.963***	0.914***	0.992***	0.839***	0.982***	0.986***	0.136	0.261	0.972***	0.846***	0.277		
Ground	0.407*	0.464*	-0.216	0.443*	0.232	0.442*	0.336	0.352*	0.428*	-0.316	-0.629	0.520**	-0.085	0.060	0.383*	
Temperature	0.885***	0.870***	0.574**	0.846***	0.887***	0.857***	0.721***	0.916***	0.859***	0.209	0.048	0.857***	0.728***	0.378*	0.878***	0.574**

Table 17. Correlation of head droopers with other parameters at Dholewala in 2001

	Droopers	Adult	Sub Adult	Juvenile	Full Droop	Half Droop	Full Crop	Half Crop	Empty Crop	Unknown	Front	Back	Lateral	Tree	Ground
Adult	0.951***														
Sub Adult	0.876***	0.862***													
Juvenile	0.980***	0.871***	0.815***												
Full Droop	0.994***	0.914***	0.882***	0.991***											
Half Droop	0.046	0.321	-0.062	-0.108	-0.065										
Full Crop	0.301	0.437*	0.569**	0.173	0.262	0.349									
Half Crop	0.919***	0.977***	0.756***	0.844***	0.870***	0.434*	0.421*								
Empty Crop	0.980***	0.888***	0.848***	0.986***	0.991***	-0.112	0.143	0.834							
Unknown	0.934***	0.934***	0.833***	0.886***	0.906***	0.239	0.555**	0.947	0.846***						
Front	0.935***	0.891***	0.977***	0.897***	0.940***	-0.056	0.546**	0.820	0.899***	0.916***					
Back	0.989***	0.947***	0.799**	0.971***	0.974***	0.118	0.210	0.937	0.968***	0.915***	0.872***				
Lateral	0.979***	0.909***	0.942***	0.962***	0.988***	-0.092	0.373	0.844	0.969***	0.907***	0.979***	0.937***			
Tree	0.989 ***	0.917***	0.836***	0.985***	0.990***	-0.027	0.161	0.876	0.996***	0.872***	0.892***	0.986***	0.963***		
Ground	0.834 ***	0.877***	0.837**	0.755***	0.799***	0.302	0.752***	0.883	0.715***	0.966***	0.895***	0.791***	0.834***	0.741***	
Temperature	0.859***	0.866***	0.857***	0.801***	0.856***	0.017	0.163	0.755	0.886***	0.680***	0.811***	0.840***	0.862***	0.884***	0.590**

Table 18. Correlation of head droopers with other parameters at Toawala in 2002

	Droopers	Adult	Sub Adult	Juvenile	Full Droop	Half Droop	Full Crop	Half Crop	Empty Crop	Front	Back	Lateral	Tree	Ground
Adult	0.995[***]													
Sub Adult	0.956[***]	0.942[***]												
Juvenile	0.919[***]	0.895[***]	0.799[***]											
Full Droop	0.962[***]	0.974[***]	0.940[***]	0.812[***]										
Half Droop	0.995[***]	0.984[***]	0.945[***]	0.941[***]	0.931[***]									
Full Crop	0.678[***]	0.607[***]	0.776[***]	0.665[***]	0.541[***]	0.716[***]								
Half Crop	0.988[***]	0.983[***]	0.962[***]	0.885[***]	0.934[***]	0.989[***]	0.705[***]							
Empty Crop	0.974[***]	0.988[***]	0.890[***]	0.890[***]	0.974[***]	0.956[***]	0.497[**]	0.944[***]						
Front	0.486[**]	0.458[**]	0.643[***]	0.298[*]	0.431[**]	0.497[**]	0.658[***]	0.583[***]	0.348[*]					
Back	0.980[***]	0.992[***]	0.920[***]	0.871[***]	0.977[***]	0.963[***]	0.549[***]	0.959[***]	0.989[***]	0.368[*]				
Lateral	0.859[***]	0.807[***]	0.841[***]	0.904[***]	0.734[***]	0.889[***]	0.866[***]	0.856[***]	0.755[***]	0.602[***]	0.742[***]			
Tree	0.998[***]	0.996[***]	0.966[***]	0.893[***]	0.971[***]	0.989[***]	0.670[***]	0.990[***]	0.973[***]	0.503[***]	0.984[***]	0.837[***]		
Ground	0.749[***]	0.709[***]	0.593[***]	0.944[***]	0.609[***]	0.785[***]	0.585[***]	0.699[***]	0.720[***]	0.172	0.670[***]	0.866[***]	0.706[***]	
Temperature	0.870[***]	0.881[***]	0.718[***]	0.895[***]	0.790[***]	0.883[***]	0.411[**]	0.869[***]	0.894[***]	0.370[*]	0.851[***]	0.761[***]	0.853[***]	0.813[***]

51

Table 19. Correlation of head droopers with other parameters at Dholewala in 2002

	Droopers	Adult	Sub Adult	Juvenile	Full Droop	Half Droop	Full Crop	Half Crop	Empty Crop	Unknown	Front	Back	Lateral	Tree	Ground
Adult	0.988***														
Sub Adult	0.845***	0.880***													
Juvenile	0.581***	0.449**	0.190												
Full Droop	0.988***	0.999***	0.873***	0.458**											
Half Droop	0.250	0.100	-0.031	0.885***	0.100										
Full Crop	0.591***	0.586***	0.257	0.423*	0.597***	0.069									
Half Crop	0.162	0.142	-0.016	0.229	0.176	-0.061	0.148								
Empty Crop	0.808***	0.854***	0.774***	0.196	0.862***	-0.199	0.743***	0.232							
Unknown	0.589***	0.481**	0.243	0.880***	0.472*	0.845***	0.417**	0.218	0.149						
Front	0.822***	0.770***	0.762***	0.635***	0.758***	0.550***	0.152	0.150	0.391*	0.742***					
Back	0.987***	0.996***	0.875***	0.469***	0.997***	0.115	0.632***	0.153	0.885***	0.484	0.750***				
Lateral	0.744***	0.681***	0.306*	0.786***	0.697***	0.429**	0.644***	0.556***	0.505***	0.615***	0.495**	0.684***			
Tree	0.991***	0.991***	0.904***	0.497	0.992***	0.169	0.538***	0.165	0.840***	0.497*	0.809***	0.991***	0.672***		
Ground	-0.333	-0.415	-0.752	0.357*	-0.415	0.466**	0.136	0.089	-0.543	0.414**	-0.239	-0.420	0.209	-0.456	
Temperature	0.809***	-0.135	-0.376	0.782***	-0.131	0.893***	-0.014	0.107	-0.443	0.699***	0.268	-0.138	0.451**	-0.097	0.744***

Table 20. Non droopers at Toawala study site in 2001

MONTH	NO. OF Observations	Percentage of Non Droopers		Percentage of Crop Status							
		No	% age	Full No	% age	Half No	% age	Empty No	% age	Unknown No	% age
JAN	481	481	100.00	86	17.87	36	7.50	328	68.19	31	6.44
FEB	879	873	99.30	109	12.49	106	12.14	642	73.53	16	1.83
MAR	432	426	98.60	30	7.04	84	19.71	306	71.83	6	1.41
APR	918	686	74.70	56	8.16	126	18.36	484	70.60	20	2.92
MAY	1,397	1,082	77.40	98	9.05	312	28.84	666	61.55	6	0.55
JUN	1,584	1,153	72.80	197	17.09	279	24.19	677	58.72	0	0.00
JUL	592	451	76.18	80	17.73	125	27.71	246	54.54	0	0.00
TOTAL	6,283	5,152	81.99	656	12.73	1,068	20.72	3,349	65.00	79	1.53

Table 21. Non droopers at Dholewala study site in 2001

MONTH	NO. OF Observations	Percentage of Non Droopers		Percentage of Crop Status							
		No	% age	Full No	% age	Half No	% age	Empty No	% age	Unknown No	% age
JAN	447	447	100.00	76	17.00	190	42.51	181	40.49	0	0.00
FEB	392	392	100.00	70	17.86	113	28.83	194	49.49	15	3.83
MAR	448	444	99.11	57	12.84	75	16.89	247	55.63	65	14.64
APR	527	478	90.70	116	24.27	153	32.01	175	36.61	34	7.11
MAY	737	536	72.73	41	7.65	162	30.22	239	44.59	94	17.54
JUN	818	649	79.34	48	7.40	185	28.51	292	44.99	124	19.11
JUL	309	246	79.61	3	1.22	83	33.74	137	55.69	23	9.35
TOTAL	3,678	3,192	86.78	411	12.87	961	30.11	1,465	45.90	355	11.12

Table 22. Non droopers at Toawala study site in 2002

MONTH	NO. OF Observations	Percentage of Non Droopers		Percentage of Crop Status							
				Full		Half		Empty		Unknown	
		No	% age	No	% age	No	% age	No	% age	No	% age
MAR	234	226	96.58	38	16.81	51	22.57	137	60.62	0	0.00
APR	864	786	90.97	126	16.03	176	22.39	484	61.58	0	0.00
MAY	844	691	81.87	113	16.35	151	21.85	425	61.51	2	0.29
JUNE	1,134	858	75.66	175	20.40	189	22.03	490	57.11	4	0.47
JULY	1,170	968	82.74	260	26.86	219	22.62	489	50.52	0	0.00
AUG	900	793	88.11	210	26.48	146	18.41	437	55.11	0	0.00
SEPT	457	423	92.56	69	16.31	82	19.39	262	61.94	10	2.36
TOTAL	**5,603**	**4,745**	**85.00**	**991**	**20.88**	**1,014**	**21.36**	**2,724**	**57.40**	**16**	**0.34**

Table 23. Non droopers at Dholewala study site in 2002

MONTH	NO. OF Observations	Percentage of Non Droopers		Percentage of Crop Status							
				Full		Half		Empty		Unknown	
		No	% age	No	% age	No	% age	No	% age	No	% age
MAR	124	96	77.42	5	2.55	32	33.33	53	55.21	6	3.06
APR	553	420	75.94	35	8.33	104	24.47	184	43.81	97	23.09
MAY	423	354	83.68	23	6.49	45	12.71	105.00	29.66	181	51.12
JUN	1,526	1,265	82.89	104	8.22	85	6.71	546	43.16	530	41.89
JUL	829	657	79.25	25	3.80	75	11.41	439	66.81	118	17.96
AUG	934	771	82.55	91	11.80	135	17.51	374	48.51	171	22.17
SEP	949	730	77.25	24	3.28	141	19.31	355	48.63	210	28.76
TOTAL	**5,338**	**4,293**	**80.42**	**307**	**7.15**	**617**	**14.37**	**2,056**	**47.89**	**1,313**	**30.58**

Table 24. Correlations of Non Droopers with crop status at Toawala 2001

	Non Droopers	Full crop	Half crop	Empty crop
Full crop	0.770^{***}			
Half crop	0.861^{***}	0.606^{***}		
Empty crop	0.964^{***}	0.662^{***}	0.724^{***}	
Unknown	-0.308	-0.303	-0.611	-0.143

Table 25. Correlations of Non Droopers with crop status at Dholewala 2001

	Non Droopers	Full crop	Half crop	Empty crop
Full crop	0.333^{*}			
Half crop	0.705^{***}	0.379^{*}		
Empty crop	0.856^{***}	-0.007	0.309	
Unknown	0.758^{***}	-0.240	0.243	0.870^{***}

Table 26. Correlations of Non Droopers with crop status at Toawala 2002

	Non Droopers	Full crop	Half crop	Empty crop
Full crop	0.906^{***}			
Half crop	0.980^{***}	0.856^{***}		
Empty crop	0.976^{***}	0.797^{***}	0.957^{***}	
Unknown	-0.308	-0.370	-0.331	-0.260^{***}

Table 27. Correlations of Non Droopers with crop status at Dholewala 2002

	Non Droopers	Full crop	Half crop	Empty crop
Full crop	0.833^{**}			
Half crop	0.515^{***}	0.439^{**}		
Empty crop	0.940^{***}	0.704^{***}	0.529^{***}	
Unknown	0.899^{***}	0.763^{***}	0.224	0.723^{***}

Table 28.Mortality of Oriental white- backed Vulture at Toawala colony in 2000-2001

Month	Total	Adult	% adult	Sub-adult	% Sub-adult	Juv/Flg	% F/J	Unknown	% unknown
DEC	3	1	33.33	0	0.00	0	0.00	2	66.66
JAN	11	8	72.72	1	9.09	2	18.18	0	0.00
FEB	3	2	66.66	1	33.33	0	0.00	0	0.00
MAR	18	9	50.00	3	16.66	6	33.33	0	0.00
APR	24	8	33.33	0	0.00	16	66.66	0	0.00
MAY	65	2	3.07	1	1.53	62	95.38	0	0.00
JUN	11	3	27.27	0	0.00	8	72.72	0	0.00
JUL	6	2	33.33	1	16.66	3	50.00	0	0.00
TOTAL	141	35	24.82	7	4.96	97	68.79	2	1.41

Table 29. Mortality of Oriental white- backed Vulture at Dholewala colony in 2000-2001

Month	Total	Adult	% adult	Sub-adult	%S-adult	Juv/Flg	% F/J	Unknown	% Unknown
NOV	3	3	100.00	0	0.00	0	0.00	0	0.00
DEC	13	13	100.00	0	0.00	0	0.00	0	0.00
JAN	10	6	60.00	1	10.00	3	30.00	0	0.00
FEB	27	18	66.66	2	7.40	7	25.00	0	0.00
MAR	11	7	63.63	3	27.27	1	9.09	0	0.00
APRIL	41	13	31.70	5	12.19	23	56.09	0	0.00
MAY	92	7	7.60	3	3.26	82	89.13	0	0.00
JUNE	24	10	41.66	3	12.50	7	29.16	4	16.66
JULY	2	1	50.00	1	50.00	0	0.00	0	0.00
Total	223	78	34.97	18	8.07	123	55.15	4	1.79

Table 30. Mortality of Oriental white- backed Vulture at Toawala colony in 2001-2002

Month	Total	Adult	% Adult	Sub-adult	% S-adult	Juv/Flg	% F/J	Unknown	% Unknown
OCT	11	6	54.54	0	0.00	0	0.00	5	45.45
NOV	2	2	100.00	0	0.00	0	0.00	0	0.00
DEC	6	4	66.66	2	33.33	0	0.00	0	0.00
JAN	7	6	85.71	1	14.28	0	0.00	0	0.00
FEB	15	8	53.33	7	46.66	0	0.00	0	0.00
MAR	31	13	41.93	12	38.70	6	19.35	0	0.00
APR	16	6	37.50	3	18.75	7	43.75	0	0.00
MAY	27	3	11.11	3	11.11	20	74.07	1	3.70
JUN	20	8	40.00	4	20.00	8	40.00	0	0.00
JUL	20	6	30.00	7	35.00	7	35.00	0	0.00
AUG	30	16	53.33	6	20.00	9	30.00	0	0.00
Total	185	78	42.16	45	24.32	57	30.81	6	3.24

Table 31. Mortality of Oriental white- backed Vulture at Dholewala colony in 2001-2002

MONTH	Total	ADULT	% Adult	SUB:ADULT	%S.Adult	JUV/FLG	% F/J	UNKNOWN	% Unknown
OCT	51	44	86.27	6	11.76	1	1.96	0	0.00
NOV	42	34	80.95	8	19.04	0	0.00	0	0.00
DEC	14	11	78.57	3	21.42	0	0.00	0	0.00
JAN	13	10	76.92	3	23.07	0	0.00	0	0.00
FEB	20	15	75.00	5	25.00	0	0.00	0	0.00
MAR	18	11	61.11	5	27.77	2	11.11	0	0.00
APR	37	7	18.91	4	10.81	26	70.27	0	0.00
MAY	53	13	24.52	5	9.43	35	66.03	0	0.00
JUN	13	7	53.84	6	46.15	0	0.00	0	0.00
JUL	34	25	73.52	1	2.94	8	23.52	0	0.00
AUG	25	21	84.00	0	0.00	4	16.00	0	0.00
Total	320	198	61.87	46	14.37	76	23.75	0	0.00

Table 32. Comparison of mortality of Oriental White- backed Vulture between Toawala and Dholewala colony in 2001 & 2002

Site	Total dead	ADULT	SUB:ADULT	JUV/FLG	UNKNOWN
Toawala 2000-2001	141	35	7	97	2
Dholewala2000-2001	223	78	18	123	4
Toawala 2001-2002	185	78	45	57	6
Dholewala 2001-2002	320	198	46	76	0
TOTAL	869	389	116	353	12

Table 33. Mortality rate of Adult and Adult + Sub adult of OWBV at Toawala and Dholewala colony in 2000 - 2001 and 2001 - 2002 breeding season

Colony	Dead Adult	Dead Adult + Sub Adult	Active nests	Mortality rate of Adult	Mortality rate of Adult+ sub Adult
Toawala 2000-2001	35	42	638	0.077	0.093
Dholewala 2000-2001	78	96	800	0.099	0.122
Toawala 2001-2002	78	123	575	0.106	0.167
Dholewala 2001-2002	198	244	589	0.262	0.323

Table 34. Gout cases found at Toawala colony in 2000-2001 breeding season (study sites and TW "Other" sites)

Month	Adult		Sub-Adult		Adult + S-Adult		Fledgling		Chick	
	Gout	No Gout	Gout	No Gout	Gout	No Gout	Gout	No Gout	Gout	No Gout
Jan	0	0	0	1	0	1	0	0	0	1
Feb	0	0	0	0	0	0	0	0	0	0
Mar	0	0	1	1	1	1	0	0	0	1
Apr	1	1	0	0	1	1	0	6	0	4
May	0	0	0	0	0	0	4	26	0	2
Jun	0	1	0	0	0	1	0	2	0	0
July	0	1	0	1	0	2	0	1	0	0
Total	1	3	1	3	2	6	4	35	0	8
Percentage	25.0	75.0	25.0	75.0	10.3	89.7	10.3	89.8	0.0	100.0

Table 35. Gout cases found at Dholewala colony in 2000-2001 breeding season (study sites and DW "Other" sites)

Month	Adult		Sub-Adult		Adult + S-Adult		Fledgling		Chick	
	Gout	No Gout	Gout	No Gout	Gout	Gout	Gout	No Gout	Gout	No gout
Dec	1	0	0	0	1	0	0	0	0	0
Jan	0	0	0	1	0	0	0	0	0	0
Feb	6	1	1	0	7	0	0	0	0	2
Mar	1	0	1	0	2	0	0	0	0	1
Apr	4	1	0	1	4	3	5	0	3	4
May	1	0	0	0	1	0	3	10	0	3
Jun	3	0	0	0	3	0	1	0	0	0
July	0	0	0	0	0	0	0	0	0	0
Total	16	2	2	2	18	3	9	10	3	10
Percentage	88.9	11.1	50.0	50.0	81.8	23.1	47.4	52.6	23.1	76.9

Table 36. Gout cases found at Toawala colony in 2001-2002 breeding season (study sites and TW "Other" sites)

Month	Adult		Sub-Adult		Adult +S-Adult		Fledgling		Chick	
	Gout	No Gout	Gout	No Gout	Gout	No Gout	Gout	No Gout	Gout	No Gout
Oct	0	1	0	0	0	1	0	0	0	0
Nov	0	0	0	0	0	0	0	0	0	0
Dec	3	1	1	1	4	2	0	0	0	0
Jan	4	0	1	0	5	0	0	0	0	0
Feb	3	0	5	0	8	0	0	0	0	0
Mar	3	1	3	0	6	1	0	0	0	0
Apr	1	0	0	0	1	0	0	1	0	0
May	2	1	1	1	3	2	2	8	0	0
Jun	7	0	3	1	10	1	5	0	0	0
July	4	1	2	2	6	3	1	0	0	0
Aug	5	0	1	0	6	0	0	0	0	0
Total	32	5	17	5	49	10	8	9	0	0
Percentage	86.5	13.5	77.3	22.7	83.1	16.9	47.1	52.9	0.00	0.00

Table 37. Gout cases found at Dholewala colony in 2001-2002 breeding season (study sites and DW "Other" sites)

Month	Adult		Sub-Adult		Adult+SubAdult		Fledgling		Chick	
	Gout	No Gout	Gout	No Gout	Gout	No Gout	Gout	No Gout	Gout	No Gout
Oct	2	2	1	1	3	3	0	0	0	0
Nov	8	1	6	1	14	2	0	0	0	0
Dec	3	1	2	0	5	1	0	0	0	0
Jan	5	0	1	0	6	0	0	0	0	0
Feb	4	0	3	0	7	0	0	0	0	0
Mar	2	0	3	0	5	0	0	0	1	0
Apr	0	0	2	0	2	0	1	2	0	0
May	5	0	3	0	8	0	1	4	3	0
Jun	3	1	0	0	3	1	0	0	0	0
July	3	0	0	0	3	0	0	0	0	0
Aug	3	0	0	0	3	0	0	0	0	0
Total	38	5	21	2	59	7	2	6	4	0
Percentage	88.4	11.6	91.3	8.7	89.4	10.6	25.0	75.0	100.0	0.0

Table 38. Necropsies of dead Oriental White- backed Vulture done at Toawala and Dholewala study sites from 2001-2002 season in Punjab Province, Pakistan

Site with Year	Total Necropsies	Gout Present	% age	Fat present	% age
Toawala 2001 Season	57	6	10.53	51	89.47
Dholewala 2001 Season	55	31	56.36	37	67.27
Toawala 2002 Season	76	56	73.68	52	68.42
Dholewala 2002 Season	78	65	83.33	60	76.92
Total Necropsies at TW & DW 2001-2002 Season	266	158	59.39	200	75.18

Table 39. Assessment of sex in Necropsies of dead Oriental White- backed Vulture done at Toawala and Dholewala study sites from 2001-2002 seasons in Punjab Province, Pakistan

Site with Year	Total # of sex asses	Male with Gout	Male without Gout	Female with Gout	Female without Gout	%age of Male with Gout	%age of Female with Gout
Toawala 2001 Season	7	0	3	1	3	0.0	14.28
Dholewala 2001 Season	12	6	4	2	0	50.0	16.66
Toawala 2002 Season	47	32	2	11	2	68.1	23.40
Dholewala 2002 Season	50	28	2	17	3	56.0	34.0
Total Assessment of sex at TW DW 2001-2002 Season	116	66	11	21	8	56.89	18.10

Table 40. Necropsies of dead Oriental White- backed Vulture done at Toawala and Dholewala study sites with their age class from 2001-2002 season in Punjab Province, Pakistan.

Site with Year	Total Necropsies	Adult	Sub-adult	Fledgling	Chick	Unknown
Toawala 2001 Season	57	4	4	39	8	2
Dholewala 2001 Season	55	18	4	19	13	1
Toawala 2002 Season	76	37	22	17	0	0
Dholewala 2002 Season	78	43	23	8	4	0
Total vulture at TW & DW 2001-2002 Season	266	102	53	83	25	3
%age	100	38.34	19.92	31.20	9.39	1.13

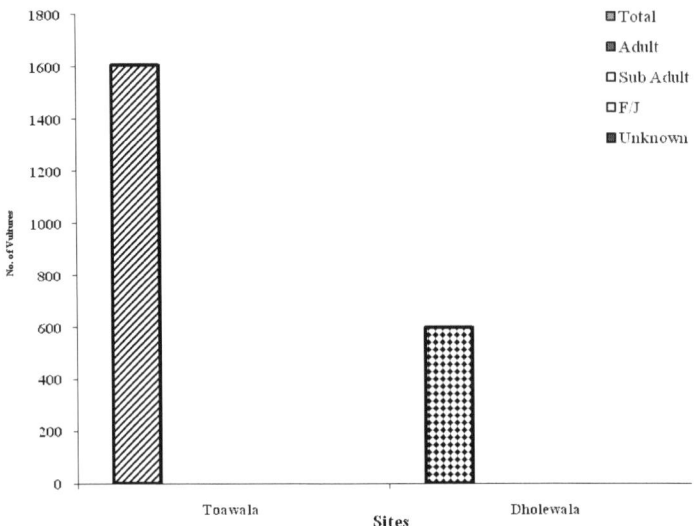

Fig. 3:Comparison of OWBV population at Dholewala and Toawala whole sites in 2001

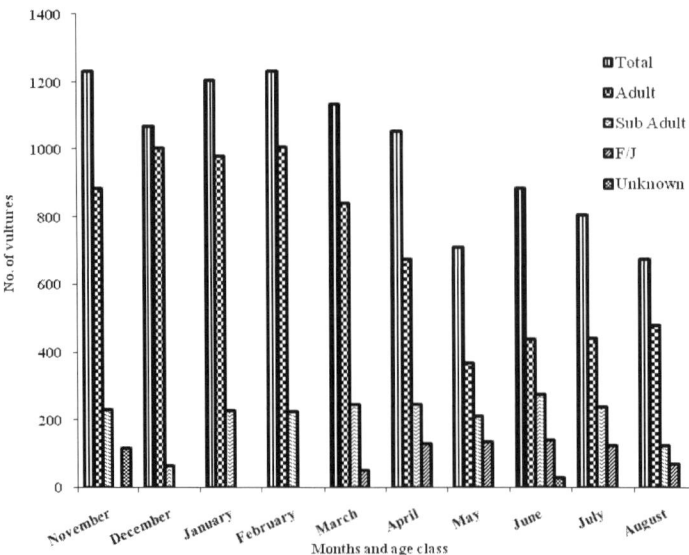

Fig. 4:Population of OWB Vultures at Toawala whole site in 2002

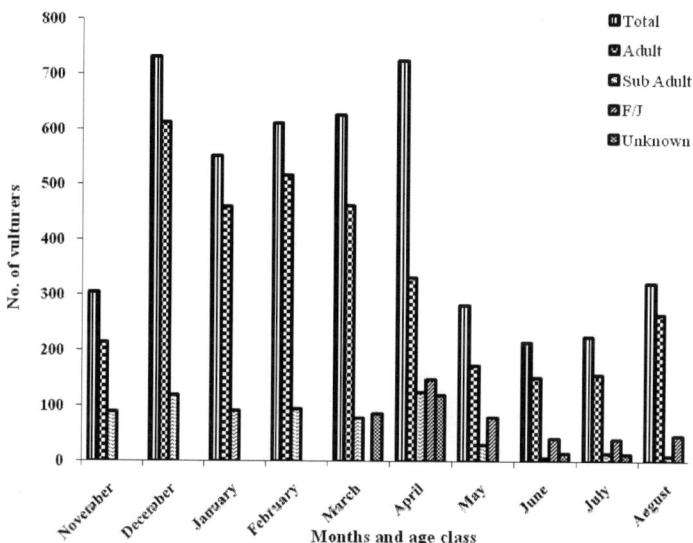

Fig. 5: Population of OWB Vultures at Dholewala whole
site in 2002

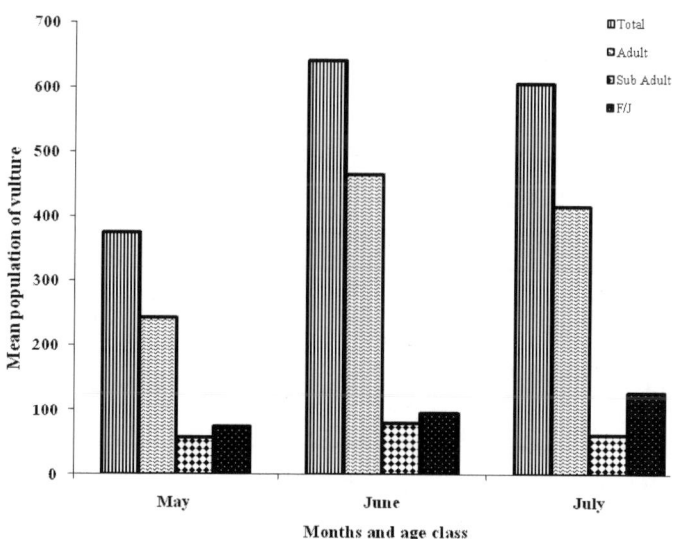

Fig. 6: Population of OWBV in weekly intensive transects
at Toawala in 2001

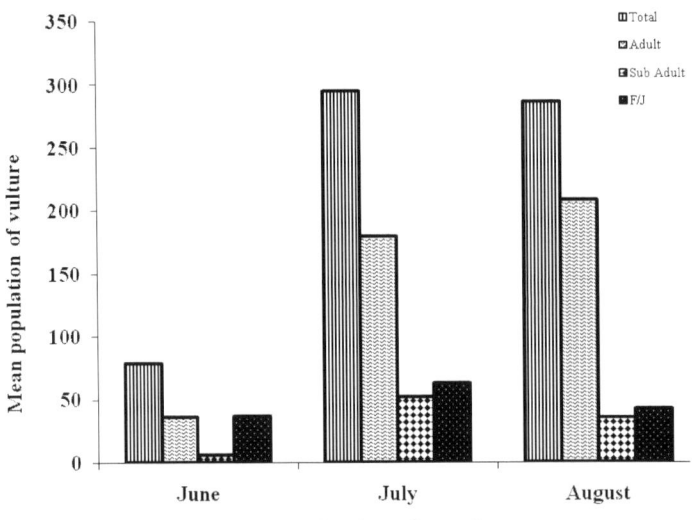

Fig. 7: Population of OWBV in weekly intensive transect
at Dholewala in 2001

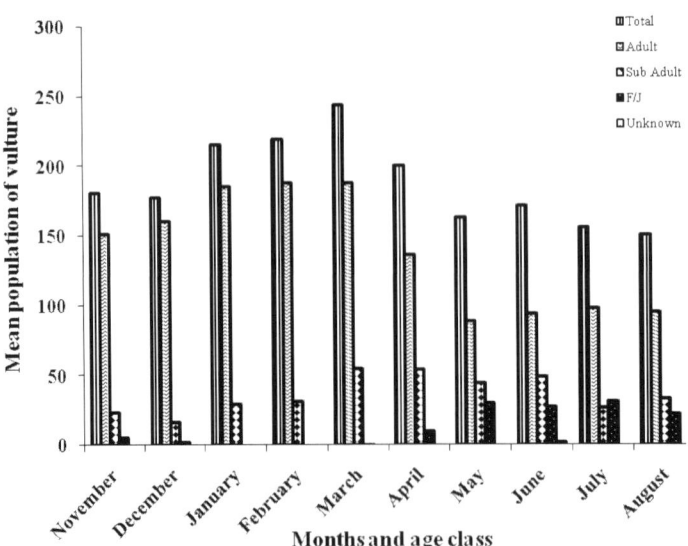

Fig. 8: Population of OWBV in weekly intensive transects
at Toawala in 2002

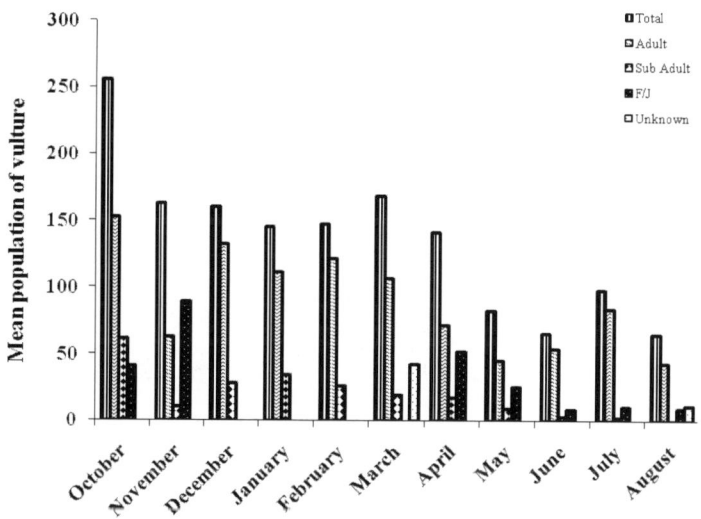

Months and age class
Fig. 9: Population of OWBV in weekly intensive transects at Dholewala in 2002

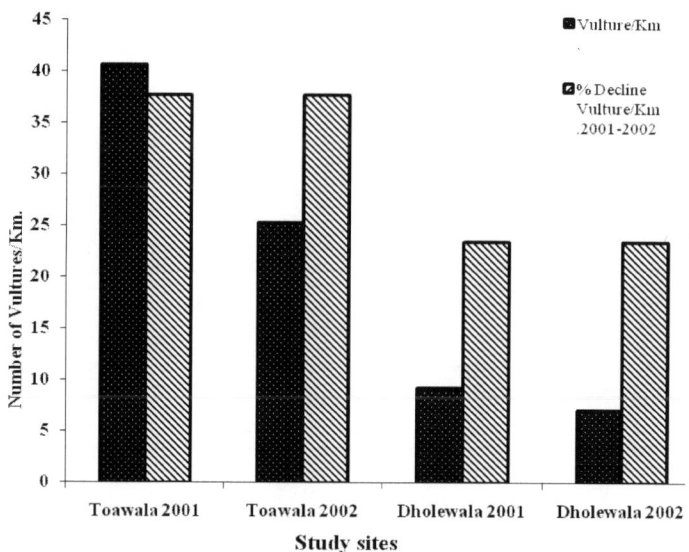

Study sites

Fig.10.% Decline vultures/Km. at Toawala and Dholewala in whole site from 2001-2002

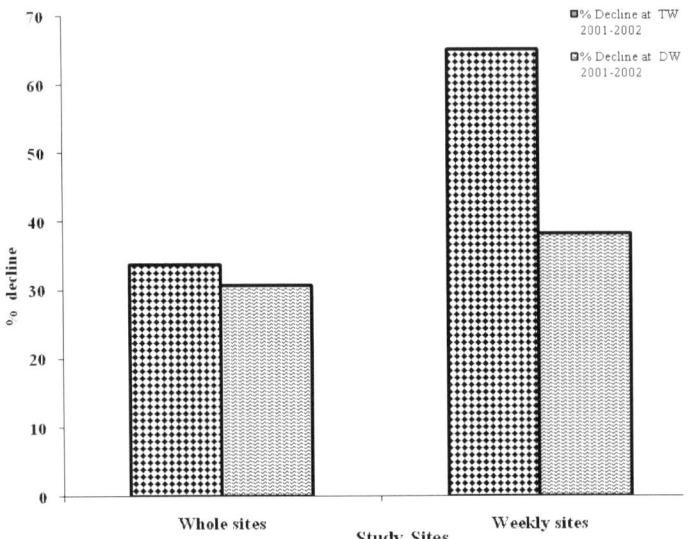

Fig 11. % Decline of Vultures at Toawala & Dholewala sites
from 2001-2002

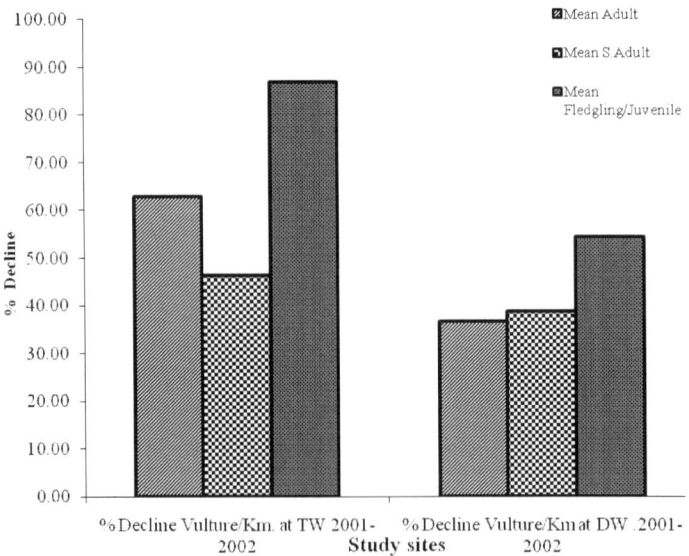

Fig 12. % Decline Vultures/ Km at Toawala and Dholewala in
Intensive transects from 2001-2002

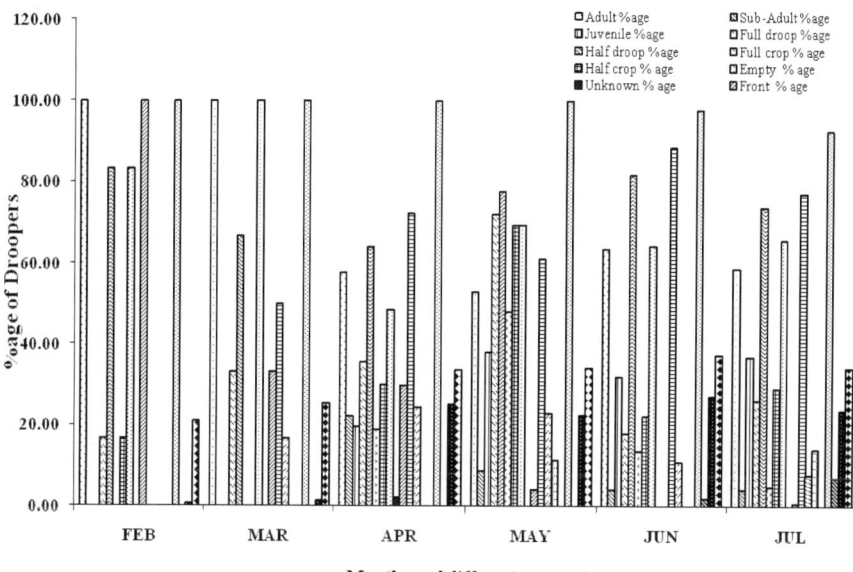

Fig. 13: Head drooping with different parameters at Toawala site in 2001

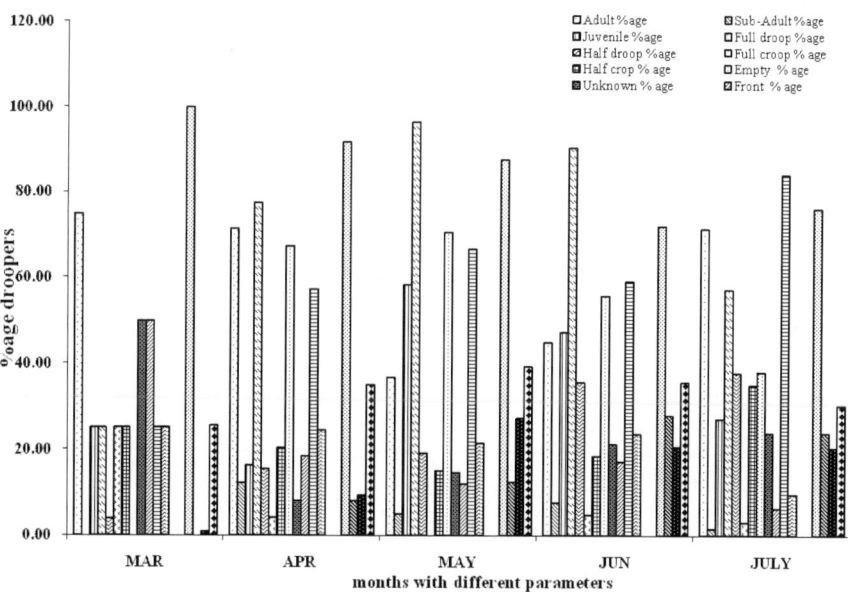

Fig. 14: Head drooping with different parameters at Dholewala site in 2001

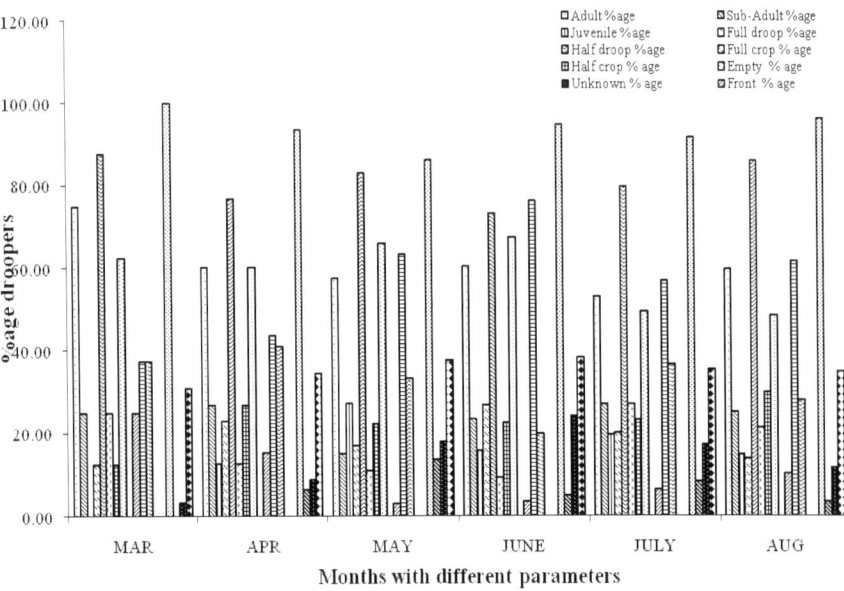

Fig. 15: Head drooping with different parameters at Toawala site in 2002

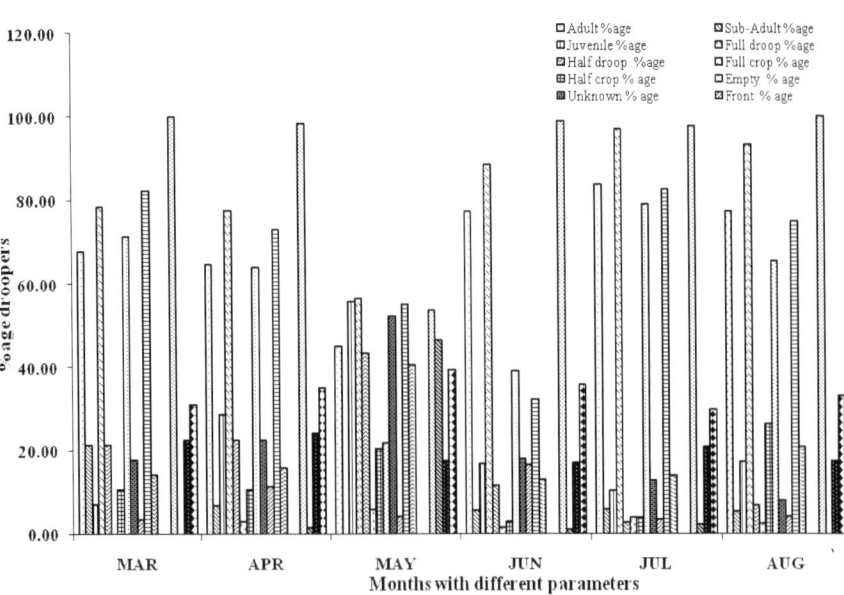

Fig. 16: Head drooping with different parameters at Dholewala site in 2002

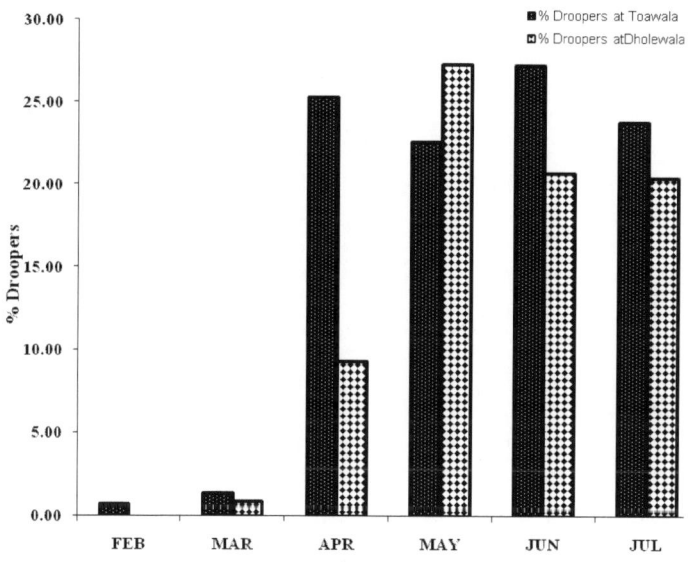

Months
Fig. 17: Droopers at Toawala and Dholewala in 2001

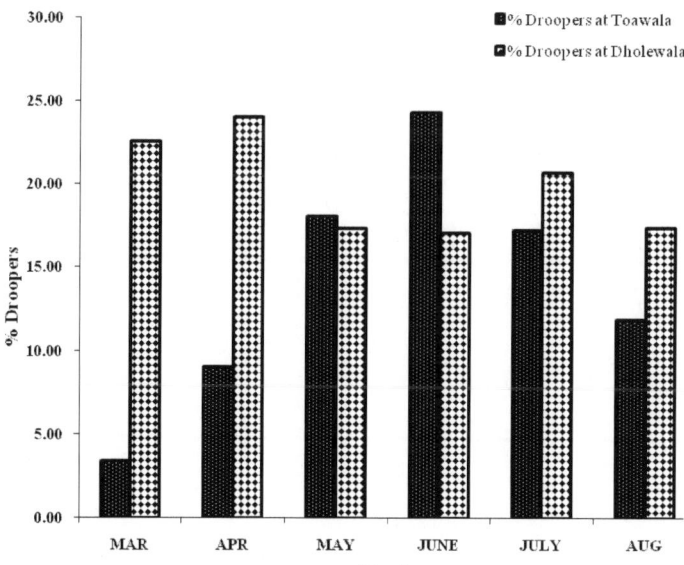

Months
Fig. 18: Droopers at Toawala and Dholewala in 2002

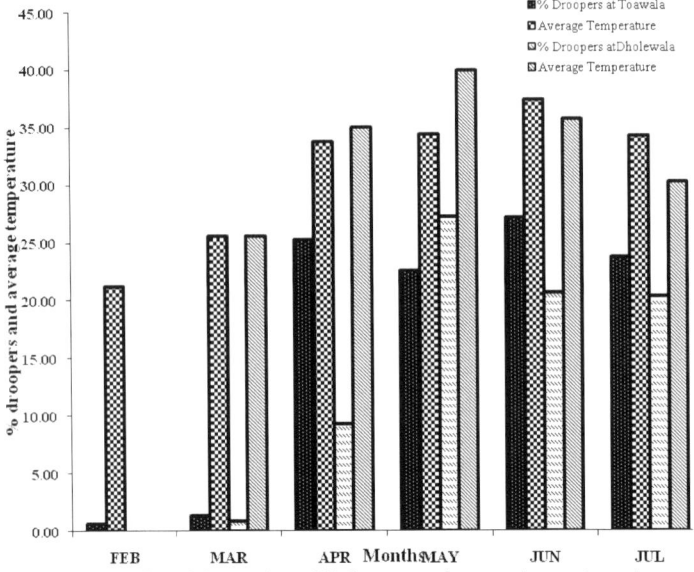

Fig. 19:Comparison of % droopers and average temperature at
Toawala and Dholewala in 2001

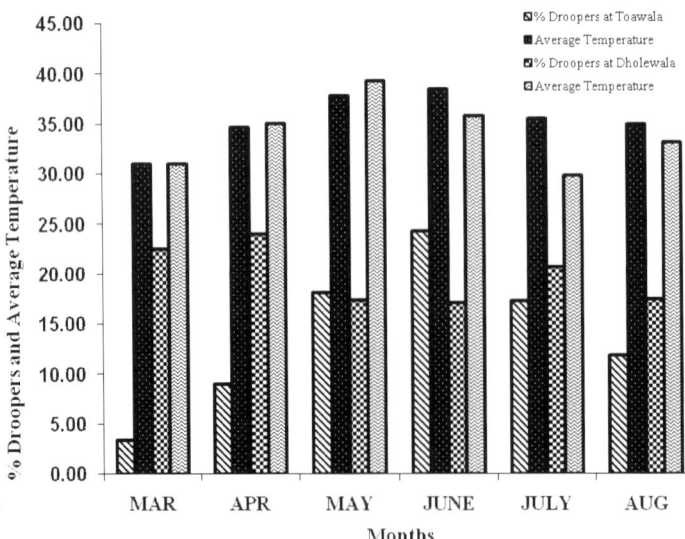

Fig. 20:Comparison of % Droopers and average
Temperature at Toawala and Dholewala in 2002

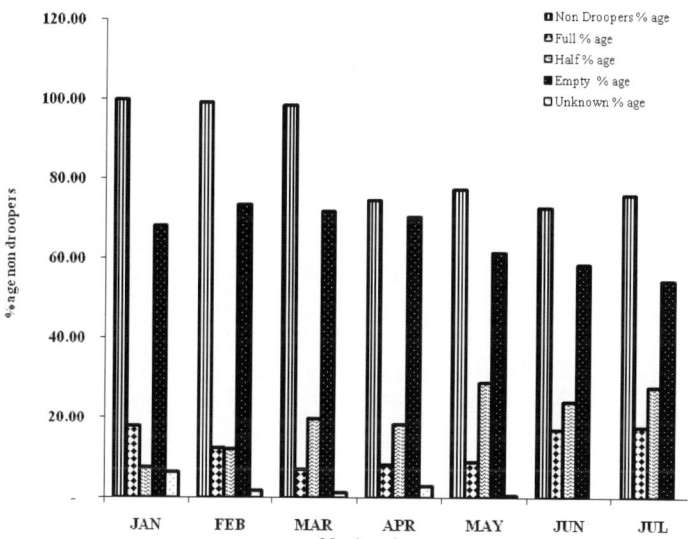

fig. 21:Crop status of %age non droopers at Toawala site in
2001

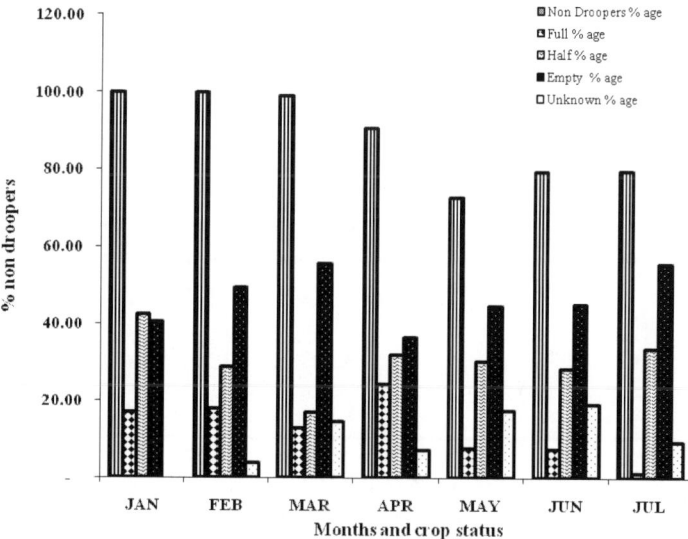

Fig. 22:Crop status of % non droopers at Dholewala site in
2001

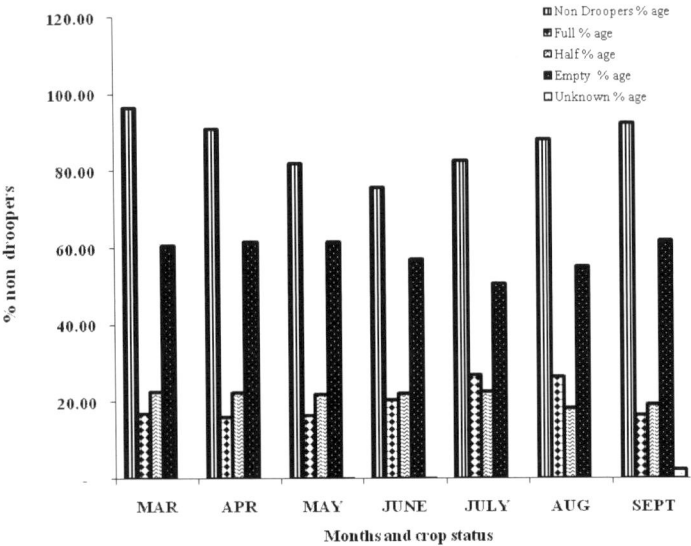

fig. 23:Crop status of % non droopers at Toawala site in 2002

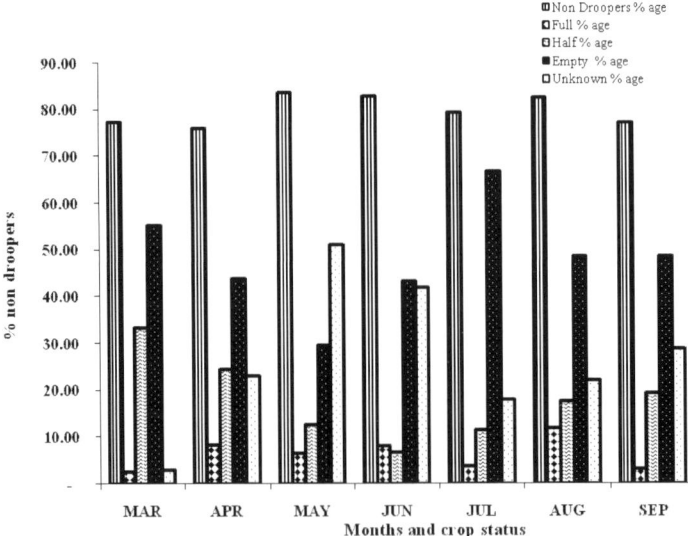

Fig. 24: Crop status of % non droopers at Dholewala site in 2002

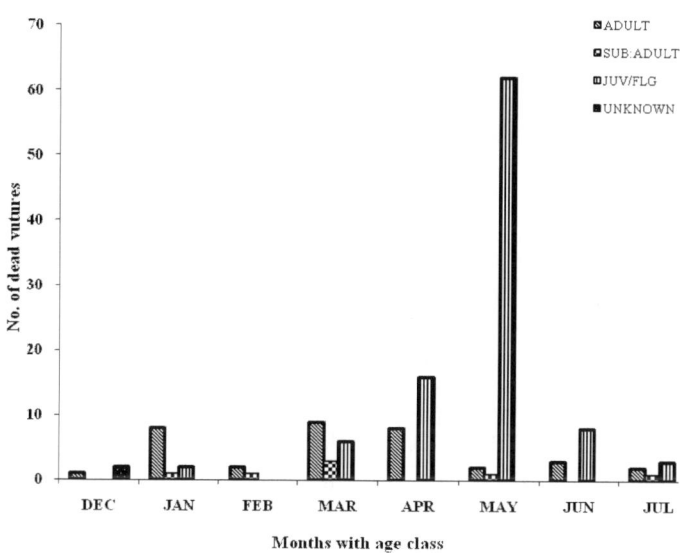

Fig. 25: Mortality of OWBV at Toawala colony in 2000-2001

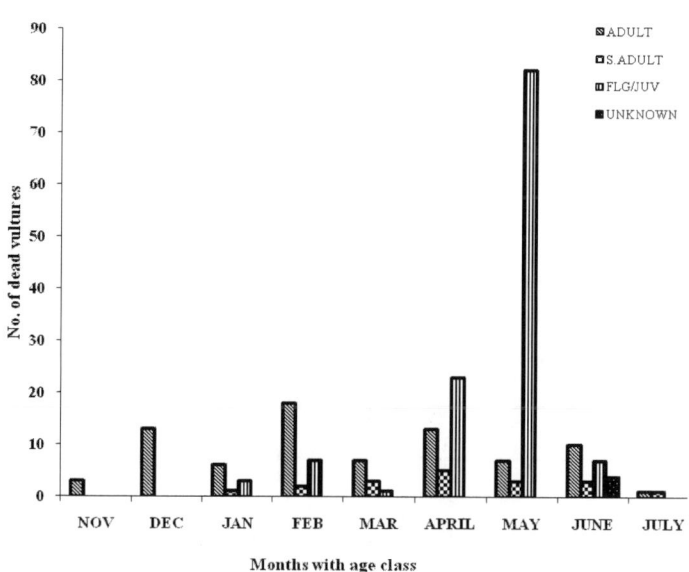

Fig. 26: Mortality of OWBV at Dholewala colony in 2000-2001

72

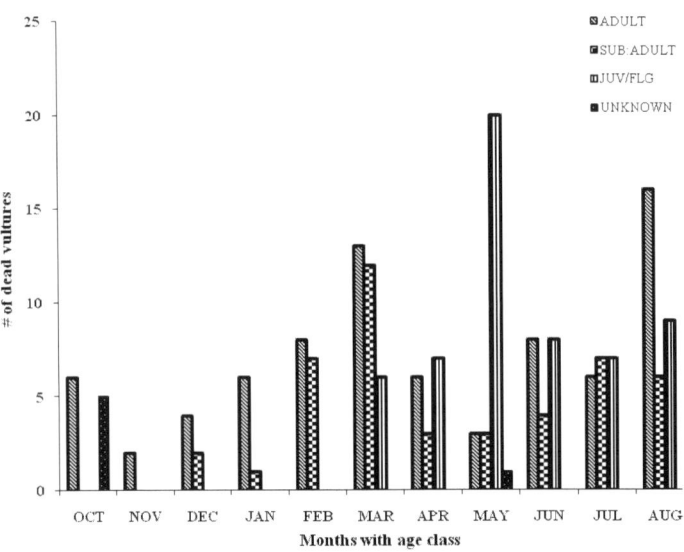

Fig. 27: Mortality of OWBV at Toawala colony in 2001-2002

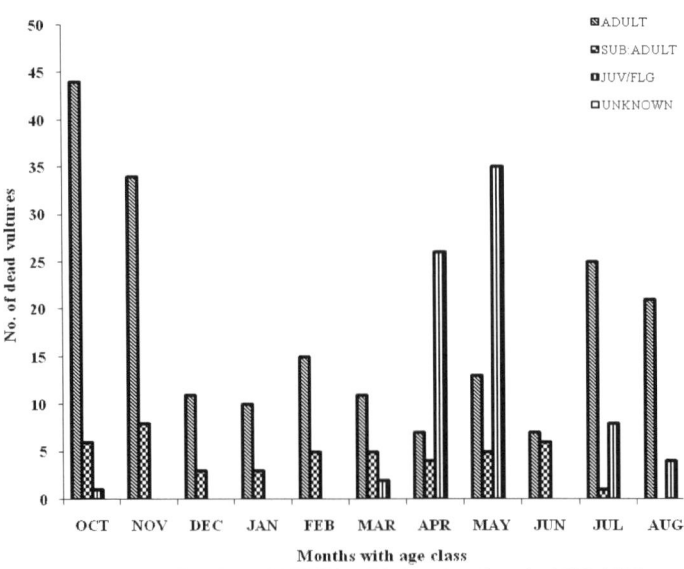

Fig. 28: Mortality of OWBV at Dholewala colony in 2001-2002

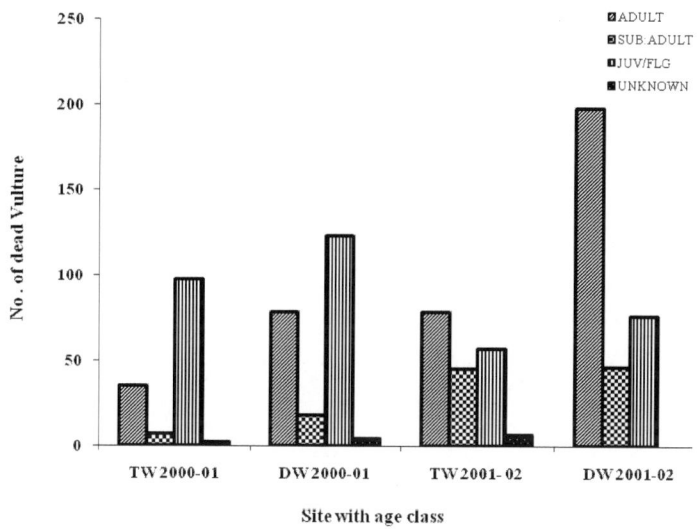

Fig. 29: Comparison of mortality of OWBV between Toawala
and Dholewala sites in 2001-2002

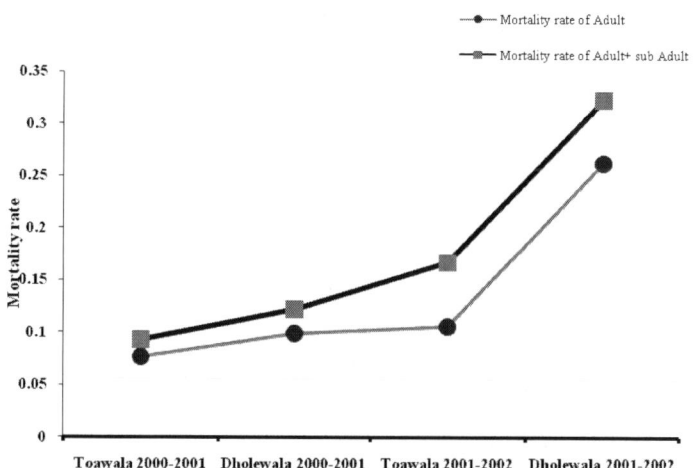

Fig.30: Comparison of mortality rate in OWBV at Toawala and
Dholewala Sites from 200-2001 &2001-2002

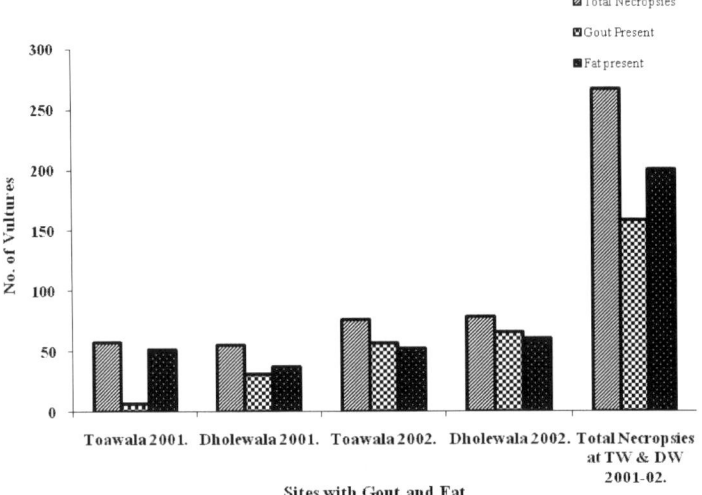

**Fig. 31: Comparison of Gout and Fat present in OWBV at
Toawala and Dholewala sites from 2001-02**

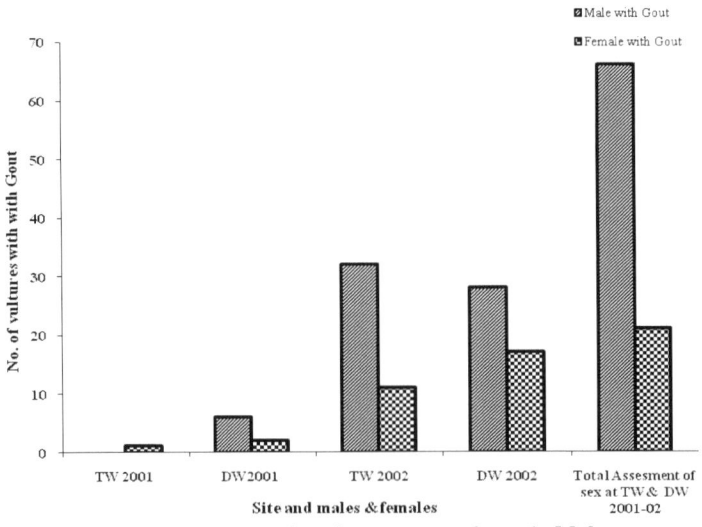

**Fig. 32: Comparison of total assessment of gout in Males
and Females of OWBV at Toawala and Dholewala sites in
2000-02**

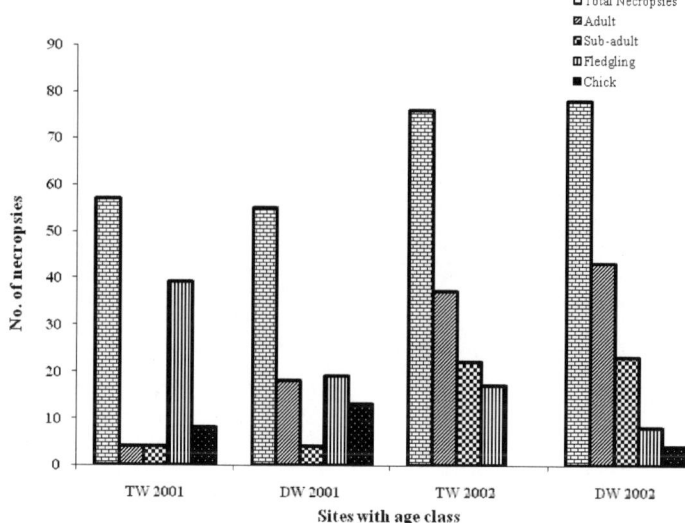

Fig. 33: Comparison of age class in necropsies at Toawala
and Dholewala in 2001-02

CHAPTER -06

DISCUSSION

This study was conducted to investigate the population decline of vultures including head drooping behaviour, mortality rate and pattern of mortality in two vulture colonies, (i.e. Toawala and Dholewala) in Punjab Province, Pakistan.

The adult to juvenile ratio recorded for White-backed Vulture was 9:1, which probably suggests breeding failure in the species as they raise only one chick every year and they have high 'breeding success. The ratio in a normal population is close to one juvenile per pair of adults (Prakash and Rahmani 2000a).

Total population at study sites

In present study in 2001 breeding season, a maximum population of Oriental White-backed Vulture at Toawala and Dholewala colony was counted as 1607 and 600 respectively (Table 1). While in second breeding season in 2002 from November 2001 to August 2002 a mean population of 1000 and 459 were observed at Toawala and Dholewala Colony respectively (Table 2).

Total birds population at Toawala having adults 71.17%, sub adult 20.90%, fledgling 1.46%, while unknown aged groups were 1.46% (Table 32). Whereas, at Dholewala Colony adults 72.84%, sub adults 14.35%, fledglings and Juvenile 7.73%, while unknown aged groups were 5.07% (Table 3).

In intensive transects at Toawala Colony in 2001, a sum of 69.38% adult, 12.33% sub adults and fledglings/Juveniles as 18.27% were found (Table 4). On the other hand, at Dholewala Colony during 2001, 64.26% adults, 14.15% Sub adults and 21.57% fledglings/juveniles were found (Table 5). In 2002 at Toawala, adults were 37.79%, sub adults were 19.32%, fledglings/juveniles were 6.87% while unknown vultures were 0.09% (Table 6).Whereas at Dholewala in 2002, and adults were 66.07%, sub adults were14.32%, fledglings/juveniles were 15.83% while unknown vultures were 3.60% (Table 7).

A total of 4,500 vultures (mainly this species) were found at 24 carcass-skinning centers in Delhi in the early 1980s; an average of 2,045 individuals (maximum 4, 0100) were seen at Timarput, Delhi, during December 1981 to July 1982, but only 612 were seen on average (maximum 1,500) in 1982 (Ali and Grubh 1984); and 3,000 vultures used to frequent a bone mill at Dasna, Delhi. In Delhi in the early 1970 s, about 400 pairs of White–backed Vulture were found (Galushin 1971). Satheesan (1999) found OWBV 10-20 years ago in the city.

The population of the White-backed Vulture crashed by over 92% in protected areas. No vulture was sighted in Gir National Park although Grubh (1974) estimated a, population of 456 vultures during the month of June during 1970-71 (Grub 1983, 1989a). Similarly, no vultures were recorded in Little Rann of Kutch where a population of 375 vultures was estimated during the survey in 1990-93. Only 135 White-backed Vultures were sighted in Desert National Park. Only 20% juveniles were recorded, whereas the adult population was 56%. The rest were sub adults. Along the highways in western India also the population has plummeted all over (Prakash and Rahmani 2000).

Khan A.A. (2001) observed population of Oriental White-backed Vultures at Dholewala 200 vultures, at Kundian forest 300 vultures, at Changa Manga more than 500 vultures, in Lahore city 31 vultures, in Wazirabad 105 vultures, at Head Islam 21 vultures, while at Dinganullah only 20 Oriental White-backed Vultures were observed.

Khan A.A. (2001) observed 52.4% sub-adults, while adult and Juveniles comprises 23.8% each.

Vulture density at study sites

However, Vulture density per Kilometer at Toawala was 40.68 vultures / Km in 2001 and 25.32 vultures / Km in 2002. However, at Dholewala, density of vultures was 9.36 vultures / Km in 2001 and 7.16 Vultures / Km in 2002 (Table 10).

In intensive transects, birds found at Toawala colony in 2001 were 70.53 vulture / Km. While at Dholewala colony, 43.13 vultures/ Km. Similarly in 2002 season, vultures / Km noted at Toawala were 24.51 vultures / Km. While at Dholewala in 2002, about 26.66 vultures / Km were noted (Table 10).

Pataudi areas (20 km^2) in Gurgaon District, Punjab 19 pairs of White-backed Vultures (*Gyps bengalensis*), 4 pairs of Egyptian Vultures (*Neophron percnopterus*) were noted. Half of the nests were in villages or within 200 m of villages. Total average density was 13.5 pairs per 10 km^2. Rohtak area (20 km^2) also in Punjab but 70 km north-west of Delhi where 36 pairs of White-backed and 4 pairs of Egyptian Vultures were noted. A density was 26.5 pairs per 10 km^2 (Galushin 1971).

The abundance of Long-billed Vultures was highest in protected areas at one vulture per 5.2 km, followed by adjoining areas (one vulture per 13.6 km) and along highways (one vulture per 22.2 km).The abundance of Long-billed was also highest in protected areas at one vulture per 6.48 km, followed by areas adjoining protected areas, one vulture in 21.47 km and along highways, one vulture per 91.59 km. (Prakash 2000).

Population decline at study sites

While in recent study, 33.77% decline in vultures population was observed from 2001 to 2002 at Toawala. Whereas 30.71% decline was observed during the same season at Dholewala. While in intensive study transects decline was 65.17% and 38.21% at Toawala and Dholewala Colony respectively (Table 9).

Additionally, at Toawala Colony % decline in vultures / Kilometer area was 37.75% whereas at Dholewala, the decline was 23.50% in a year.

In conclusion during 2001 - 2002, both at Toawala and Dholewala, substantial decline in population of *Gyps bengalensis* was noted.

The recent vulture population declines may have commenced in the early 1990s, but they only became visible at some locations in the mid-1990s, and became particularly serious after 1997 (Satheesan 2000a). General declines of "vultures" were reported: "all over northern India" in the late 1990s, with many newspaper reports commenting on carcasses being left uneaten because of the scarcity of vultures (Rahmani 1998a). Such declines have been noted in north-west India, where total numbers of all vulture species "often exceeded 1,000 per 100 km" along roads between Karachi and Delhi in the 1980s (Thiollay 2000).

Populations of two commonest griffon vultures (genus *Gyps)* have declined by >90% during the last decade. Both affected species, the White-backed and Long-billed Vultures *(Gyps bengalensis* and *G. indicus),* were once regarded as very common in India, but now they are listed as critically endangered by the IUCN. At Bharatpur 96% decline in Long-billed Vulture and a 97% decline in Long-billed Vultures between 1985 and 1999 were observed (Prakash 1999).

The population of White-backed appears to have crashed in the east zone. A 91 % decline in the population was recorded in protected areas. Similarly a sharp decline in population of White-backed was observed along highways where it declined by more than 99% (Prakash and Rahmani 2000b).The population of Long-billed also crashed by about 99% (Prakash and Rahmani 2000). The population declined by more than 86% from 1990 – 1993 (Prakash and Rahmani 2000).

In 1996 the number of Vultures declined 25-30% and Kites continued their slight decline (on 10-20%) (Galushin and Zakharova-Kubareva 1998).

High adult mortality and a corresponding decrease in numbers of active nests throughout the study sites indicate a population decline. Findings suggest that mortality rates differ between both colonies, and this has been supported by a corresponding variation in population decline. This implies that the mortality factor affects vultures at the both sites at different rates. The decline of vultures in India has coincided with an almost three-fold increase in the use of pesticides in the region over the last decade. Previously common resident raptors such as White-eyed Buzzards *Butastur teesa* and Black-shouldered Kites *Elanus caeruleus* are now rarely encountered in the

Punjab Province, suggesting that *Gyps* vultures may not be the only genus in decline. Temporal and spatial clusters of dead vultures have been located indicating a point source of exposure.

Head drooping behaviour

Protracted 'neck-drooping' appears to be a sign that birds are weak or clinically sick possibly carried out to conserve energy. It is every important behaviour to monitor, as it is the only obvious behavioural indication that birds are ill, and even where this is reported in healthy birds under hot conditions, it is likely that it will be recorded more frequently in populations with a higher proportion of sick or weak birds. As with many non-specific signs of sickness, this may be exacerbated by environmental conditions, such as excessive temperature, and in future surveys it is important that temperature and time of day are recorded (Gilbert *et al.* 2002).

Similarly, in present study a total of 18,194 Oriental White-backed Vultures were observed at two study sites (Toawala and Dholewala) in Punjab Province Pakistan for the purpose of head drooping analysis, a sum of 3,267 (17.95%) vultures displayed head drooping. Out of these vultures that displayed head drooping, adults were 1,990 (60.91%), sub adults were 376 (11.50%) and juvenile/fledglings were 900 (27.54%). Out of these drooping vultures, 1,582 (48.42%) birds displayed full drooping (as their head touching the claws in shade) while 1,685 (51.57%) displayed half drooping intensity. Crop status was observed as full crop in 294 (8.99%), half full crop in 668 (20.44%), empty crop was comparatively high in 1,958 (59.93%) and unknown crop status was 254 (7.77%) shown in (Table 11).These results indicate that majority of head drooping vultures had empty crops, which suggests the posture may help conserve energy. We suggest that head drooping in vultures in the Indian subcontinent may be a previously overlooked normal behavioral response to increasing ambient temperature and possibly other stresses. Thus, all head drooping vultures are not necessarily sick but sick vultures may droops their heads to reduce body temperature, conserve energy and alleviate stress related to sickness.

Head drooping occurred in vultures while they are perched either on trees or on the ground: their heads and necks drooped almost to the point of touching their feet, giving a sickly and lethargic impression.

Head drooping was characteristically observed in groups of vultures in which mortality was occurring. Heads drooped downwards, as though the birds were resting, touching the ground if the bird was perched on the ground. After assuming this position, birds could, however, recover, fly, feed, and feed young and otherwise assume normal behaviour; five deaths were recorded of birds that showed this behaviour over a period of 30 days (Prakash 1999).

In Keoladeo National Park (Prakash 1999), the birds were usually found dead on the nest, on trees or on the ground below the trees. Several deaths were observed. Prior to death, individual vultures were seen perched on trees, dozing, with the neck slowly limping down and would wake up with a start, as the beak hit the branch. The bird usually remained in this condition for more than 30 days (n=5) and then would fall off the branch, sometimes getting caught in the branches of the trees and at times falling on the ground. The birds would die within minutes of falling (Prakash 2000).The ability of the birds showing this behaviour to raise their heads, to fly, and to feed rules out a hypothesis that it is due to a neurological or muscular problem. Rather, it appears to be a symptom of general lethargy of birds that are sick, becoming more severe as the disease progresses (Cunningham 2000).

Droopers comparison with Drooping intensity

While in recent study 1,131 (19.434%) and 486 (17.12%) vultures displayed head drooping at Toawala and Dholewala respectively in 2001 and 824 (16.01%) and 826 (18.81%) vultures displayed head drooping at Toawala and Dholewala Colony in 2002.

Drooping intensity was also recorded with full drooping vultures as 271 (23.96%) and 422 (86.83%), while half droopers as 860 (76.03%) and 64 (13.16%) at Toawala and Dholewala respectively in 2001 (Table 12, Table 13). While in 2002, Drooping intensity was observed as full droopers 175 (21.23%) and 714 (86.44%) and half droopers as 649 (78.76%) and 112 (13.55%) at Toawala and Dholewala respectively (Table14, Table 15).

Head drooping was observed in15.6% vultures at Changa Manga, in Lahore city 22.6%, at Head Islam 33.3%, at Dinganullah15%, while in Lal Suhanra National Park 20% of Oriental White-backed Vultures were observed in head drooping posture (A.A. Khan 2001).

At Koshi Tappu Wildlife Reserve one dead individual was found during a visit in 2000 and about 15-20% of individuals presumably including or mainly comprising, OWBV were observed showing the head-drooping behaviour noted in India; in Royal Chitwan National Park no vultures were seen (Rahmani and Prakash 2000). Head drooping behaviour was noted in 17% of individuals of White-rumped Vultures during surveys by BNHS in April-June 2000 in north and central India (Prakash 2000).

Over 50% showed the head drooping syndrome characteristic of sick vultures (Prakash 2000). A higher proportion of *Gyps bengalensis* (44%) had neck droop in the North (Pain *et al.* 2003). Robertson (1986) found 53% head droopers with empty crop.

Head drooping behaviour has been reported in other species of *Gyps* vultures, but under somewhat different circumstances. Mundy *et al.* (1992) refer to "dozing behaviour" of African White-backed vultures *Gyps africanus*; during the day birds may doze with their necks drooping, almost touching their feet.

Head drooping is also observed in Cape vultures *Gyps coprotheres* and African White-backed vultures after feeding during hot weather, with heads occasionally touching the ground (Verdoorn 2000).

Droopers comparison with Temperature

While in recent study, mean temperature was also recorded for droopers and it was found 31.11 Celsius and 33.19 Celsius at Toawala and Dholewala colony in 2001 (Table 12, Table 13) and 35.40 Celsius and 34.02 Celsius at Toawala and Dholewala study sites respectively in 2002 (Table14, Table 15).

While head drooping is highly positively correlated with the temperature ($r = 0.885$, $p<0.001$) at Toawala in 2001 and ($p<0.001$, $r = 0.870$) at Toawala in 2002.

It is clear that head drooping is highly positively correlated with the temperature ($P<0.001$, $r = 0.859$) at Dholewala in 2001 and ($p<0.001$, $r = 0.809$) at Dholewala in 2002.

The neck drooping behaviour is usually displayed while they are resting during hot hours of the day. The highest numbers of sick vultures, perched with drooping necks, were seen in Desert National Park and on the border of Rajasthan and Gujarat (Prakash and Rahmani 2000).

Virani *et al*. (2001) reported a correlation between neck drooping and ambient temperature in a population of Oriental White-backed Vultures in Pakistan, where large numbers of vultures are also reported to be dying. A strong positive correlation between the proportion of head drooping vultures and increasing ambient temperatures was observed.

Droopers comparison with age class

In present study it was further found that 670 (59.23%) and 233 (47.94%) vultures were adult, sub adult as 104 (9.19%) and 30 (6.17%) and juvenile / fledgling were 357 (31.56%) and 223 (45.88%) at Toawala and Dholewala respectively, in 2001 (Table 12, Table 13). While in 2002, 479 (58.13%) and 608 (73.60%) vultures were adult, sub adult as 193 (23.42%) and 49 (5.93%) while juvenile and fledglings were 152 (18.44%) and 168 (20.33%) at Toawala and Dholewala study sites respectively (Table14, Table 15).

In the Koshi Tappo Wildlife Reserve in Nepal, where a total of 128 White-backed and 2 Slender-billed vultures was recorded, at least 16 of the White-backed Vultures showed the head drooping syndrome; 68.75% (11) were immature, 12.5% (2) were juveniles and 18.75% (3) were adults (Prakash *et al*. 2003).

Droopers comparison with orientation to sun

Orientation of the birds were also taken into account and it was observed that vultures facing to sun were 31 (2.74%) and 68 (13.99%) while backing to sun were exceedingly higher 855 (75.59%) and 316 (65.02%), lateral to sun direction were189 (16.71%) and 10 (20.98%), while sun above to vultures were observed in 56 (4.95%) and 0 (0%) at Toawala and Dholewala study sites respectively in 2001 (Table 12, Table 13). While orientation to the sun was also recorded for each vulture in 2002 and it was found that vultures facing to sun were 53 (6.43%) and 75 (9.07%), vultures backing to sun were exceedingly higher as 526 (63.83%) and 606 (73.36%) and lateral to sun direction were 245 (29.73%) and 145 (17.55%) at Toawala and Dholewala respectively (Table14, Table 15).

The fact that nearly two-thirds of the vultures observed in this study with drooping heads had their backs to the sun, are results consistent with a behavioral response to reduce heat load. It is suggested that by shading the highly vascularized skin on the head and neck, vultures may be able to direct blood to this surface to lose heat.

Droopers comparison with Crop status

Crop status was also noted for droopers, as full crop was 125 (11.05%) and 13 (2.67%) vultures, half filled crop was 288 (25.46%) and 94 (19.34%), empty crop vultures were exceedingly higher 713 (63.04%) and 293 (60.28%) while unknown were 5 (0.04%) and 86 (17.69%) at Toawala and Dholewala respectively in 2001 (Table 12, Table 13). While crop status was also observed in 2002 and it was found that vultures with full crop were 133 (16.14%) and 23 (2.78%), half filled crop were 197 (23.90%) and 89 (10.77%) and empty crop vultures were exceedingly higher 491 (59.58%) and 461 (55.81%). Crop status of some vultures were not observed as 0 (0%) and 153 (18.52%) at Toawala and Dholewala colony respectively (Table14, Table 15).

Similarly crop status of non droopers was observed and it was found that high number of vultures had empty crop as observed in droopers at both sites in 2001 and in 2002.

Droopers comparison with perch substrate

Parching substrate for all vultures displayed drooping was also taken into account and it was observed in 1,112 (98.32%) and 395 (81.27%) vultures on tree, while rest 19 (1.67%) and 91 (18.72%) were on ground in Toawala and Dholewala sites respectively in 2001 (Table 12, Table 13). While Perched substrate was also observed in 2002, as 763 (92.59%) and 785 (95.03%) vultures were on trees while on ground it was observed in 61 (7.40%) and 41 (4.96%) at Toawala and Dholewala study sites respectively (Table14, Table 15).

Mortality and Morbidity

While in present study 869 dead, sick and injured Oriental White-backed Vultures were found from Toawala and Dholewala from December 2000 to August 2002. From these, 141 dead Oriental White-backed Vultures were collected and removed from Toawala in December 2000-July 2001 while 223 Vultures were collected in November 2000-July 2001 at Dholewala (Table32).

Similarly, a sum of 185 dead, sick and injured vultures were found at Toawala from October 2001 - August 2002 while a sum of 320 dead, sick and injured Vultures were found and removed in October 2000- August 2001 (Table32).

In Nepal, (Virani *et al.* 2001) collected 45 dead White backed Vultures, of which 34 (75.5%) were adults.

A method uses plumage characters to assess the age class distribution within a population and assumes the reduction in representation of each age class is due to mortality and the population is stable. This method is only useful where age class can be assessed in the field with confidence. Although it is possible to subdivide a population of Oriental White-backed Vultures into broad age classes as adult, sub adult and juvenile (Newton 1979).

Overall the proportion of individuals' head-drooping, and number of deaths reported, was highest in areas near the border with Ranjasthan, India, and lower in the Indus River areas (Rahmani and Prakash 2000). At Kundian forest, Mianwali district, "large-scale deaths" of vultures were found in April-May 2000, and no sick individuals were observed in that month (A. A. Khan 2001).

Mortality comparison with age class

While in present study 389 (44.76%) vultures were adult, 116 (13.34%) were sub adult, 353 (40.62%) were nestling/fledgling and 12 (1.38%) were Unknown or unidentified (which were found as only bones, feather or wings piles).From these, at Toawala, adults were accounted 35 (24.82%), sub adult 7 (4.96%) while nestling/fledgling mortality was exceedingly high as 97 (68.79%) and for Unidentified vultures as 2(1.41%) in December 2000 - July 2001 (Table 28).While at Dholewala in November 2000 - July 2001, 78 (34.97%) were adult, 18 (8.07%) were sub adult, 123 (55.15%) were Nestling/fledgling and 4 (1.79%) were unidentified (Table 29). Similarly at Toawala in October 2001 - August 2002, 78 (42.16%), sub-adult were 45 (24.32%), Nestling/fledglings were 57 (30.81%) and Unidentified were 6 (3.24%) in Table 30.While at Dholewala in October 2001 - August 2002, 198 (61.87%) were adult, sub adults were 46 (14.37%) and Nestling/fledglings were 76 (23.75) in Table 32.

Adult and sub adult mortality rate was consistently high during the study period while nestling mortality was low until April and May in both breeding seasons. In these months mortality of nestling increased due to fledgling fatalities. This is believed to be attributed to naivete (unable to take off after their first flight).

Old World vultures conform to a life history strategy typical of large raptors and seabirds (Wynne-Edwards 1955, Wyk *et al* 1993, Amadon 1964, Singh *et al* 1976, Piper *et al.* 1981). They are long-lived, reproduce slowly, and adult survival is high in comparison to smaller birds. Among most animals including raptors, there is a strong positive correlation between body weight and adult survival, such that heavier species live longer, and are subject to lower mortality rates (Newton 1979).

The declines were associated with high adult and juvenile mortality, in 1985 and 1986,>1700 *Gyps bengalensis* were recorded in KNP, and 14 birds (7 adult. 7 juvenile) were found dead. In 1997-1998, when the population numbered just a few hundred birds, 73 adults and 10 juveniles were found dead (Pain *et al.* 2003). 45 dead White backed

Benson (2000) stated that fledgling corresponds to one of the periods of highest mortality for all birds, although this is very difficult to quantify in vultures due to the long distances birds may cover before dying. Studies of mortality in Cape Vulture *Gyps coprotheres* based on recovery of birds banded as nestlings suggested mortality rates in excess to 83% (Piper *et al.* 1981).

Parental care in Old World vultures continues for an extended period after fledging. During the PFDP (post fledgling dependence period) juveniles must return to the nest in order to be fed by their parents (Robertson 1985). Mortality in the PFDP is largely a consequence of the inability of young vultures to return to the nest (Benson 2000).

For large, long-lived, slow reproducing raptors more than about 5% adult annual mortality is predicted to cause population decline and a small change in this rate can have a large effect on population stability (Watson 1990). The results of this study have shown that while there is still a relatively large population of Oriental White-backed Vultures in the Punjab Province of Pakistan compared to India and Nepal, there is high adult/sub-adult mortality indicative of a rapidly declining population.Vultures were found dead, 34 (75.5%) of which were adults (Virani *et al.* 2001).

Mortality comparison with mortality rate

Similarly in the Present study, if we compare the results of mortality rate, Adult annual mortality rate was calculated 7.7% at Toawala while 9.9% at Dholewala in December 2000 to July 2001.While adult annual mortality rate was calculated as 10.6% at Toawala and at Dholewala it was 26.2% which clearly indicate population decline (Table 33).

Infectious disease does not appear to be a major cause of mortality in populations of *Gyps* vultures elsewhere in the world (Benson 2000). It was clear that the vulture declines are due to exceptionally high mortality of vultures, with all age classes being affected. Also, the reproductive rate is abnormally low. This is probably due to direct and indirect effects of the high mortality Prakash (1999), the population of both White-backed and Long-billed has crashed.

While in present study, the birds were usually found dead on the nest, on trees or on the ground below the trees. Sporadic deaths were observed. It is believed that the "true" mortality estimates could be higher since we could not account for dead birds eaten dragged away by scavengers, vultures that had died away from the nesting colony, and non-breeding individuals that roosted within area. Therefore the annual adult mortality estimates from Toawala and Dholewala probably represent a minimal estimate that may be higher. Because many dead vultures may have been missed, and the cause and extent of missing them may have been different in both site,

It is believed that the high fledgling mortality in the five-week period following fledging was related to naivete of young birds resulting in deaths from premature fledging, traumatic injury, heat stress, and dehydration and starvation due to their inability to return to their nest site to be fed by the parents. This kind and level of fledgling mortality has been documented in Cape Vultures *Gyps coprotheres* (Benson 2000).

Oriental White-backed Vultures are comparable in weight to Bald Eagles *Haliaeetus leucocephalus,* and would be expected to have a similar annual adult survival rate of >0.91.Gyps vultures are generally long-lived; one captive *Gyps fulvus* lived for 37 years (Newton 1979).

The range of annual mortality rate was 11.4% (adults only) to 18.6% (adults and sub adults combined) in Dholewala and 25.4% (adults only) to 36.2% (adults and sub adults combined) in Changa Manga. Given our assumptions, we suggest the adult only mortality rate considerably underestimates the true annual mortality of the breeding population. The range of annual mortality rates we present are higher than known and calculated mortality rates in stable populations of other raptor species of comparable body weight, demonstrating that the colonies studied were in a state of rapid decline (Gilbert *et al.* 2002).

For large, long-lived, slow reproducing raptors more than about 5% adult annual mortality is predicted to cause population decline and a small change in this rate can have a large effect on population stability (Watson 1990).

Mortality comparison with visceral gout

In the present study in Pakistan we found high number of Adult and sub adults with Gout. An assessment of gout was made in 266 freshly dead Oriental White-backed Vultures through necropsies. Of these 57 and 76 necropsies were done, From these only 6 (10.53%) and 56 (73.68%) of the dead Oriental White-backed Vultures had signs of visceral gout in which 1 (25%) were adults, sub adults were 1 (25%), fledgling were 4 (10.31%) and in nestlings (n = 8) gout was not observed, while 32 (86.5%) were adults, sub adults were 17 (77.3%), fledglings were 8 (47.1%) and no nestling was observed with gout at Toawala in first and second breeding season respectively (Table 38).

Similarly 55 and 78 necropsies were done, From these 31 (56.36) and 65 (83.33%) of the dead Oriental White-backed Vultures had signs of visceral gout, in which 16 (88.9%) adults, sub adults were 2 (50.0%), fledglings were 9 (47.4%) and 3 (23.1%) were nestling, while 32 (86.5%) were adults, sub adults were 17 (77.3%), fledglings were 8 (47.1%) and no nestling was observed with gout at Dholewala in first and second breeding season respectively. Out of these 266 necropsies, a total of 158 (59.39%) dead vultures had the signs of visceral gout (Table 38).

From all above results it is clear that Gout is playing a drastic role in vultures decline and possible causes of visceral gout are as follow. The deposition of urates on serous surfaces of thoraco-abdominal organs is termed as visceral gout. It is not a disease condition but a result/sequalae of renal failure.

Visceral gout is a sign of renal insufficiency due to primary renal disease or secondary to conditions that cause dehydration (Oaks *et al.* 2004).

The possibility of a disease specific to *Gyps* vultures in the Indian subcontinent, of which gout results as a secondary effect, seems tenable. Secondary poisoning of vultures from pesticides or carcass poisoning is a possible cause, while not impossible, has been suggested as extremely unlikely (Oaks *et al.* 2001).

Visceral gout was also recorded in juveniles and nestlings, although at a lower rate than was observed in adults and sub adults i.e.19% and 13%, respectively. This may be due to a lower susceptibility to the condition than is seen in adult and sub adult birds, a lower exposure to factors leading to visceral gout, or a low prevalence of gout cases relative to the incidence of other age-related mortality factors. It is likely that data on the occurrence of gout in nestlings has suffered from bias, with an over-representation of nestlings that have fallen from their nests. Comparison with other species of *Gyps* vultures suggests that high fledgling mortality may not be unusual for the species (Piper *et al.* 1981, Robertson 1985, Benson 2000).

Although visceral gout was only found in a minority of juveniles and nestlings, the fact that it was found, at all, should be considered significant. If the hypothesis that gout related deaths are connected is correct, then the presence of the condition in juveniles and nestlings would suggest that the mortality factor responsible for the condition is either vertically or horizontally transmissible from adults to offspring, or is inherently present in the vicinity of the colony itself (Gilbert *et al.* 2002).

The findings of a high prevalence of visceral gout in freshly dead adult/sub-adult vultures in Pakistan signifies that they are likely to be dying of the same cause that has extirpated populations of vultures in India. The fact that some nestlings and fledglings were affected by gout indicates that the cause may be transmitted from adults to progeny in the form of relay contamination at the nest, brought back to the nest in food, or may be inherently present at nest sites. The lower proportion of nestlings and fledglings affected by gout does not necessarily mean that adult and sub-adult vultures are more vulnerable to this mortality factor than younger birds. The rate of fledging mortality due to naivete was very high, unique to this age-class, and therefore, may obscure mortality due to gout because fledglings die from other causes related to naivete before they die of gout. Proportional death rates are therefore not comparable.

The prolonged pattern of unusually high adult/sub-adult mortality rates throughout the study period, the widespread prevalence of gout-affected adult and sub-adult vulture carcasses, and the low breeding success, confirms that the ultimate cause of gout-related vulture mortalities is unprecedented, regionally widespread, and rapidly reducing the numbers of Oriental White-backed Vultures in Pakistan.

The ultimate cause of gout-related mortality in populations of Oriental White-backed Vultures still remains indeterminate (Anderson and Mundy 2001).

Autopsies on vultures performed so far have found "degenerative changes in the urinary tubules" in the kidneys, and whitish deposits presumed to be urates present in the heart, liver, kidney and spleen; no evidence of bacterial infection was found (Risebrough 1999). The progressive accumulation of uric acid (as found in the human condition commonly termed gout) may therefore explain the pattern of reproductive failure and chronic condition eventually resulting in death, as observed in vultures in Keoladeo National Park (Risebrough 1999). Seven dead vultures were found and cause of death appeared to be dehydration caused by enteritis (Prakash 2000). All autopsies carried out so far on vultures found dead in the wild have shown symptoms of acute enteritis, degeneration of kidney cells, and extensive visceral gout, whilst vultures which were captured as sick individuals that died in captivity did not show visceral gout (Rahmani and Prakash 2000).

Pesticides although the toxic effect of pesticides has been implicated in population declines of a number of raptor species (Ratcliffe 1967, Hickey and Anderson 1968), and this has been suggested as an explanation of the population crashes in *Gyps* vultures in the Indian region (Rahmani 1998b, Ghatak 1999, Prakash 1999).

There has been no evidence of starvation being a contributing factor to the death of vultures necropsied from across India and Pakistan (Gilbert *et al.* 2002, Prakash *et al.* 2003).

Postmortem analyses of *G. bengalensis, G. indicus,* and one *G.himalayensis* from India (Cunningham *et al.* 2001, Pain *et al.* 2002) and *G. bengalensis* from Pakistan (Oaks *et al.* 2001, Gilbert *et al.* 2002) identified renal and visceral gout (crystallization of uric acid in the tissues) in the majority of birds found dead, and enteritis in a high proportion of the birds from India (Cunningham *et al.* 2001). Few other gross findings are consistently observed. The presence of visceral gout in tissues of dead birds from both countries supports the hypothesis that the same mortality factor is responsible for all these deaths. Although renal gout is often attributed to kidney disease, in these cases the gout was acute suggesting that this condition is a consequence of the primary disease and not the disease itself (Cunningham *et al.* 2001). Visceral gout and enteritis are non specific lesions and could result from, for example, a contaminant insult or an infectious disease process.

Histological analyses of tissues, however, found higher than expected proportions of vultures with inflammation of blood vessel walls and proliferation in the brain of glial cells (Cunningham *et al.* 2001).

At post-mortem examination, many birds showed evidence of visceral or renal gout, and similar findings have been reported from dead vultures in Pakistan. This, and the presence of similar clinical signs of sickness, such as neck-drooping, makes it highly probable that the same factor is affecting *Gyps* species irrespective of location (Oaks *et al.* 2001).

Birds are uricotelic animals i.e. end product of protein metabolism is uric acid. If any agent damages kidneys, there is loss of renal tubular function. Uric acid accumulates in blood circulation resulting in to hyperuricacidaemia, which may lead to visceral gout. Uric acid being sparingly soluble in water easily precipitates on serous surfaces. Different agents causing renal damage may be (Mishra, S.K and Prasad, G. Unpublished report).

Excess if calcium in diet (calcium nephropathy) Normal levels required are 0.6% and it produces renal lesions at 3% Ca levels. Sodium intoxication (sodium chloride, sodium bicarbonate, sodium citrate, sodium acetate, sodium monoglutamate) cause Na-K imbalance. Vitamin-A deficiency (Baby chick disease) Vitamin- A is essential for maintenance of structural and functional integrity of epithelial cells. It causes metaplasia and hyprekeratinization of ureteral epithelium (Mishra, S.K and Prasad, G. Unpublished report).

Infectious causes includes Infectious bronchitis virus (nephrotoxic 'T' strain) Nephritis nephrosis syndrome virus Infectious monocytosis/Blue comb disease Infectious bursal disease Estern equine encephalitis (Mishra, S.K and Prasad, G. Unpublished report).

Other potentional causes includes toxic chemicals, insecticides, their metabolites, poisons and toxins that damages kidneys-ochratoxin, oosporein. Starvation, inclement weather, dehydration, obstruction of ureters, due to back pressure of uroliths, cyst, tumor or abscess and thereby, causing visceral gout. The affected ureters become dilated filled with pasty white mass (Mishra, S.K and Prasad, G. Unpublished report).

Decline in other raptor populations, high mortality in other birds (Samant *et al.* 1995) and the extent of pesticide use and changes in livestock management practices and extensive use of veterinary drugs may cause a serious hazards and renal failure.

Fat comparison with necropsy

In present study, an assessment of fat was made in 266 freshly dead Oriental White-backed Vultures through necropsies. Out of these 266 necropsies, omental fat reserves were found intact in 200 (75.18%) of the dead Oriental White-backed Vultures. Of these 57 and 76 necropsies were done at Toawala in which omental fat reserves were found intact in 51 (89.47%) vultures and 52 (68.42%) in first and second breeding season respectively. While 55 and 78 necropsies were done at Dholewala in which omental fat reserves were found intact in 37 (67.27%) vultures and 60 (76.92%) in first and second breeding season respectively. All other dead vultures without showing gout were died from dehydration, heat stress, injury and other causes presumed to relate to the naivete of the birds during the post fledgling period (Table 38).

The body condition scoring system used by Houston (1976) has been greatly simplified for use in this study. Houston's system defines nine classes of body condition based on deposits of fat in the mesentery, omentum and subcutaneous tissue. In this study, birdswere scored with moderate to abundant omental fat reserves as `positive' for omental fat, which would correspond to a body condition score of 6-9 on Houston's scale. Therefore, it should not be assumed that birds scored as `negative' in our study are completely emaciated, but simply indicate that these birds have a body condition that would be considered moderate to poor (Houston 1990).

Moderate to abundant reserves of omental fat were found in 81% of dead birds. This supports the assessment of Prakash (1999) that the observed mortality in *Gyps* vultures is unrelated to the availability of food. Prakash also states that prior to death vultures would remain immobile for "more than 30 days". It is clear from this study that this is not the case for the majority of vultures that ultimately die. Moderate to abundant levels of omental fat are unlikely to be found in a bird that remains immobile for a period in excess of one month, and would suggest that death in the majority of cases in this study followed a more acute clinical course.

The affected carcasses examined so far have been in varying states of nutritional fitness, with some of the birds having substantial amounts of body fat. This is probably an indicator of the period of time which has elapsed between an individual bird becoming sick and its death – a time which is likely to have a degree of variance (Cunningham *et al.* 2001).

Above results clearly indicates that there was not a problem of food scarcity because 75% vultures had omental fat reserves intact.

Conclusion

In conclusion, this study supports the hypothesis that the mortality factor responsible for the decline in vulture populations in India is also present in Punjab Province, Pakistan.

It is clear that at both sites high numbers of adult and sub adults are found dead, which clearly indicate the population decline in the Oriental White-backed Vulture and all above mentioned results are comparable with the other authors. The high numbers of mortalities in fledgling were due to naiveté of the fledglings and this was observed in fledgling in first few weeks. The results of first and second year have shown that while there is still a large population of White-backed Vultures in the Punjab province of Pakistan (compared to India and Nepal), there is high adult mortality indicative of a rapidly declining population. The presence of visceral gout in adults and sub adults vultures, necropsied in Pakistan signifies that vultures are likely to be dying of the same cause that has extirpated vultures in India.

Mass vulture deaths reported during the summer in parts of India and Pakistan are more likely the result of fledgling deaths that occur naturally soon after the fledging period which coincides with the hot months of April and May. This period also coincides with a peak in "head drooping" behavior, which in these results suggest was a previously overlooked behavioral response to increasing ambient temperature and possibly other stresses. Thus, all head drooping vultures are not necessarily sick but sick vultures may droop their heads.

In 75% of the dead birds fats are intact so the death of vultures are sudden and found in clusters, that leads more towards relay intoxication through food resources.

RECOMMENDATIONS

Vultures are keystone species and their declines are having adverse effects upon other wildlife, domestic animals and humans. In particular, there is a risk of increase in diseases that threaten human life and welfare.

Vultures are sampling the environment and their deaths and population collapse have demonstrated a widespread toxic effect. These declines are important to conservationists and should lead to better role in the conservation of Oriental White-backed Vultures. The extinction of vultures may have far reaching economic, ecological and public health implications. Vultures may play a role in the control of important livestock diseases (e.g., anthrax, tuberculosis, brucellosis, foot and mouth disease, rinderpest and contagious pleuropneumonia) by rapid disposal of infected animals and inactivation of pathogens.

Recently Oaks *et al.* (2004) described that, the non-steroidal anti-inflammatory drug (NSAID), diclofenac, is the primary cause for the catastrophic collapse of Oriental White-backed Vultures *Gyps bengalensis* in Pakistan. This discovery raises significant global concerns about environmental contamination by drugs and has important conservation implications for *Gyps* vultures and perhaps other scavenging birds occurring in South Asia.

Remedial action to prevent vulture extinctions is needed.

1. Continued monitoring will be required to test the hypothesis that observed mortality rate will lead to a reduction in the breeding population in future.

2. There is a need to continue the diagnostic work already underway should be considered of utmost importance.

3. Priority should be given to identifying the mortality factor and understanding the manner of its interaction with the vulture population, with the ultimate aim of reducing exposure or increasing resistance of the remaining wild birds, sufficient to allow the population to recover.

4. There is need for captive breeding of Oriental White-backed Vulture at one hand and to control the use of drug and poison/contamination agents from the food of OWBV in the wild on the other hand.

5. There is need for further research on poison/contaminants involved in the decline of Oriental White-backed Vulture population in the wild, so that ecofriendly drugs/poisons should be used for agricultural and livestock management purposes.

6. Species must be upgraded in appendix-1 of CITES schedule, due to its imminent extinction rate.

7. Provisioning clean food to vultures in the wild may help reduce rates of mortality below levels that cause population decline. Vulture restaurants have been used to augment food supplies to vultures in South Africa and Europe with some success, but they have never been used to entirely substitute for an otherwise abundant source of food.

There are three species restoration scenarios that could be considered. These methods are

➤ Translocation and release of vultures from locations where vultures remain in larger numbers.

➤ Collection, holding and further release of vultures once the reason for decline is removed.

➤ Collection of vultures and establishment of captive breeding populations from which young are produced and released.

CHAPTER -07

References

Ali, S. and Ripley, S. D. (1968) *Handbook of the birds of India and Pakistan together with those of Nepal, Bhutan and Ceylon*. Vol I. Divers to Hawks. Oxford University Press. pp: 301-310.

Ali, S. and Ripley, S. D. (1983) *Handbook of the Birds of India and Pakistan*: Compact Edition. Delhi, Oxford University press, Oxford, New York.

Amadon, D. (1964) The evolution of low reproductive rates in birds. *Evolution* 18: 105-110.

Benson, P.C. (2000) Causes of Cape Vulture *Gyps coprotheres* mortality at the Kransberg colony: a 17 year update. *Raptors at Risk. Proceedings of the V World Conference on Birds of Prey and Owls*. R.D. Chancellor and B.-U. Meyburg [Eds.]. Midrand Johannesburg, South Africa 4-11 August 1998.

BirdLife International. (2001) *Threatened Birds of Asia; The BirdLife International Red Data Book*. Cambridge. pp.588-613.

Bishop, K. D. (1999) The birds and mammals recorded on the 1999 VENT Bhutan Tour. Unpublished.

Brooke, R.K. (1984). *South African Red Data Book - Birds*. South African National Scientific Programmes Report No 97.

Brown, L. and Amadon, D. (1968) *Eagles, hawks and falcons of the world*. London: Country Life Books.

Cheng Tso-hsin (1987) *A synopsis of the avifauna of China*. Beijing: Science Press.

Cunningham, A. Investigation of Vulture Mortality in India. Report of a visit to Bharatpur from 14th to 23rd February 2000 and Mumbai from 24th February to 5th March 2000, on behalf of the RSPB and BNHS.

Cunningham, A., Prakash, V., Ghalsasi, G.R. and Pain, D. (2001) Investigating the cause of catastrophic declines in Asian Griffon Vultures, *Gyps indicus and G. bengalensis*. Reports from the workshop on Indian *Gyps* Vultures. Katzner, T. & Parry-Jones, J. [Eds.]. 4th Eurasian Congress on Raptors Sevilla-Spain. September 2001.

Evans, L. B. and Piper, S. (1981) Bone abnormalities in Cape Vulture (Gyps coprotheres). *J. S. African Vet. Assoc.* 52: 67-68.

Galushin, V. M. (1971) A huge urban population of birds of prey in Delhi, India (preliminary note). *This* 113: 522.

Galushin, V. M. and Zakharova, K.A. (1998) Nesting raptor populations within urban and agricultural habitats in Northern-Central India. Asian Raptor Research and Conservation. First symposium on raptors in Asia. *Shiga,* Japan, P30.

Ghatak, A. R. (1999) What's eating the vulture? *Downn to Earth* (15 January 1999).

Gilbert, M., Virani, M. Z., Watson, R. T., Oaks, J. L., Benson, P. C., Khan, A. A., Ahmed, S., Chaudhry, J., Arshad, M., Mahmood, S. and Shah, Q. A. (2002) Breeding and mortality of White-backed vulture *Gyps bengalensis* in Punjab province, Pakistan. *Bird conservation international.* (12): 311- 326.

Gooders, J. (2000) John Gooders on life's idiosyncracies. Bird Watching August.

Grubh R. B. (1989a). Ecological study of Bird Hazard at Indian Aerodromes: Phase-II. Final Report 1982-88. Bombay Natural History Society.

Grubh R. B. (1989b). The ecology and behaviour of vultures in Gir forest Ph. D. Thesis. University of Bombay, India.

Grubh, R. B. (1974) *The bird of Gir forests* (The ecology and behaviour of vultures in Gir forest). Ph.D. thesis, Bombay University, India.

Grubh, R. B. (1983) The status of vultures in the Indian Subcontinent. Pp. 107-112 in S. R. Wilbur and J. A. Jackson, eds. *Vulture biology and management.* University of California Press, Berkeley and Los Angeles.

Hickey, J. J. and Anderson, D. W. (1968) Chlorinated hydrocarbon and eggshell changes in raptorial and fish-eating birds. *Science* 1 62: 271-273.

Hollom, P. A. D., Porter, R. F., Christensen, S. and Willis, I. (1988) *Birds of the Middle East and North Africa.* Calton, UK: T. & A. D. Poyser.

Houston, D. C. (1976) Breeding of the White-backed and Ruppell's Griffon Vultures *Gyps africanus* and *G. rueppellh. Ibis* 118: 14-38.

Houston, D. C. (1990) *The use of vultures to dispose of human corpses in India and Tibet.* P156 in I. Newton and P Olsen, eds. *Birds of prey.* London: Merehurst Press.

http//www.wetlands.org.

Inskipp, C. and Inskipp, T. P. (1991) *A guide to the birds of Nepal.* London: A. & C. Black/Christopher Helm and Washington: Smithsonian Institution Press.

Khan, A. A., Virani, M. Z., Oaks, J. L., Benson, P. C., Gilbert, M., Watson, R. T. and Risebrough, R. W. (2001) A survey of the Oriental White-backed Vulture (*Gyps bengalensis*) in the Punjab province, Pakistan. *Journal of research (science),* Bahauddin Zakariya University, Multan, Pakistan 12 (1) pp. 97-104.

Lumeij, J. T. (1994) Nephrology. Pp. 538-555 in: Ritchie.

MacKinnon, J. and Phillipps, K. (1993) *A field guide to the birds of Borneo, Sumatra, Java and Bali*. Oxford University Press.

Medway, L. and Wells, D. R. (1976) *The birds of the Malay Peninsula*, 5. London and Kuala Lumpur: H. F. and G. Witherby in association with Penerbit Universiti Malaya.

Mishra, S.K and Prasad, G. Investigations on vulture mortality. College of veterinary sciences, C. C. S. H. A. U., Hisar 125004 (Unpublished report).

Mundy, P. J. (1982). The comparative biology of southern African vultures. Johannesburg: Vulture Study Group.

Mundy, P., Butchart, D., Ledger, J. & Piper, S. (1992) *The Vultures of Africa*. London and San Diego: Academic Press.

Newton, I. (1979) *Population ecology of raptors*. Berkhamstead, UK: T. and A. D. Poyser. Newton, P N., Breeden, S. and Norman.

Oaks J.L. Diagnostic investigation into vulture mortality, Punjab Province, Pakistan www.peregrinefund.org/conserv_vulture_results.html *Nov* (2000).

Oaks, J. L., Gilbert, M., Virani, M. Z., Watson, R. T., Meteyer, C. U., Rideout, B. A., Shivaprasad, H. L., Ahmad, S., Chaudhry, M. J. I., Arshad, M., Mahmood, S., Ali, A. and Khan, A. A. (2004) Diclofenac residues as the cause of population decline of vultures in Pakistan. *Nature* 427: 630-633.

Oaks, L., and B.A. Rideout, M. Gibert, R. Watson, m. Virani, and A. Ahmed Khan. (2001) Summary of Diagnostic investigation into vulture mortality: Punjab Province, Pakistan, 2000-2001. Reports from the workshop on Indian Gyps vultures. Pages 12-13 in T. Katzner, and j. Parry-Jones, editors. Proceedings of the 4th Eurasian congress on raptors. Estacion Biologica Donana, Raptor Research Foundation, Seville.

Pain, D.J. (2001) Conservation of critically endangered Gyps vultures in India. In: Pakistan, 2000-2001. Reports from the workshop on Indian *Gyps* Vultures.

Pain, D.J., A.A. Cunningham, Donald, P.F., Duckworth, J.C., Houston, D.C., Katzner, T., Parry-Jones, J., Pool, C., V. Prakash., Round, P and Timmins, R. (2003) Causes and effect of temporospatial decline of *Gyps* Vulture in Asia. Page 661-669 in issue in international conservation. Conservation Biology, Volume17, No3, June 2003.

Pain, D.J., V. Prakash, A.A. Cunningham, and G. R. Ghalsasi. (2002) Vulture declines in India: patterns, causes and spread. Pages 4-7 in T. Katzner and J. Parry-Jones, editors. Conservation of Gyps Vulture in Asia. 3rd North American Ornithological Congress, September 24-28, 2002. New Orleans, LA, USA.

Paludan, K. (1959) On the birds of Afghanistan. *Vidensk. Medd. Dansk Naturhist. Foren.* 122:1-332.

Piper, S. E., Mundy, P. J. and Ledger, J. A. (1981) Estimates of survival in the Cape Vulture *Gyps corprotheres. Journal of animal ecology* 50: 815-825.

Prakash, V (1989) Population and distribution of raptors in Keoladeo National Park, Bharatpur, India. Pp.129-137 in B.-U. Meyburg and R. D. Chancellor, eds. Raptors in the modern world. Berlin, London and Paris: World Working Group on Birds of Prey and Owls.

Prakash, V. (1999) Status of vultures in Keoladeo National Park. Baharatpur, Rajasthan with special reference to population crash in *Gyps* species *J. Bombay Nat. Hist. Soc.* 96: 365-378.

Prakash, V. (2000). Effect of Environmental Contamination on Raptors. Annual Report 1999-2000, Bombay Natural History Society, Mumbai.

Prakash, V. and Rahmani, A. R. (2000) Notes about the decline of Indian vultures, with particular reference to Keoladeo National Park. *Vulture News* 41: 6-13.

Prakash. V., Pain, D. J., Cunningham, A. A., Donald, P. F., Prakash, N., Verma, A., Gargi, R., Sivakumar, S. and Rahmani, A. R. (2003) Catastrophic collapse of Indian White-backed *Gyps bengalensis* and Long-billed *Gyps indicus* Vulture populations. *Biological Conservation* 109:381-390.

Rahmani, A. & Prakash, V. (2000) *A brief report on the international seminar on vulture situation in India, BNHS, New Delhi 18*$^{.h}$ *to 20*$^{.h}$ *September 2000.* Mumbai: Bombay.

Rahmani, A. (1999) Vulture alert. Internet: www.wild.allindia.com/vulalert.

Rahmani, A. R. (1996) Status of vultures in the Thar desert of India. *Vulture News* 35: 25-30.

Rahmani, A. R. (1998a) A possible decline of vultures in India. *Oriental Bird Club Bull.* 28: 40-41.

Rahmani, A. R. (1998b) A possible decline of vultures in India. *Oriental Bird Club Bull.* 28: 40-41.

Rahmani, A. R. and Prakash, V. (2000a) Technical note on the catastrophic decline in vulture populations in India. Unpublished.

Rahmani, A. R. and Prakash, V. (2000b) A brief report on the International Seminar on Vulture Situation in India, organised by *Bombay Natural History Society*, 18-20 September 2000. Unpublished.

Rao, K. M. (1992) Vultures endangered in Guntur and Prakasam districts (A.P) and vulture-eating community. *Newsletter for Birdwatchers* 32: 6-7.

Rasmussen, P. C. and Parry S. J. (2000) On the specific distinctness of the Himalayan Long billed Vulture *Gyps indicus ,tenuirostris.* Abstract, 118'h Stated Meeting of the American Ornithologists' Union, Memorial University of Newfoundland, St. John's.

Risebrough R.W., Virani M.A. Threats to the Vultures of Asia, the Middle East, Europe & Africa, and strategies for an international plan. *Raptor News, Oct* (2000).

Risebrough, R. W. (1999) Population crash of the Gyps vultures in India: evidence for a disease factor and recommendations for emergency efforts. Unpublished.

Robertson, A.S. (1985) Observations on the post-fledging dependence period of Cape vultures. *Ostrich* 56: 58-66.

Rosalind, L. (2000) White-backed Vultures in the throes of an unidentified malady in the Balaram Ambaji Wildlife Sanctuary in Banaskentha district, Gujarat. *Newsletter for Birdwatchers* 40(3): 38-39.

Samant, J. S., Prakash, V. and Naoroji, R. (1995) Ecology and behaviour of resident raptors with special reference to endangered species. Final report 1990-1993. Bombay: Bombay Natural History Society.

Satheesan, S. M. (1998) The role of vultures in the disposal of human corpses in India and Tibet. *Vulture News* 39: 32-33.

Satheesan, S. M. (1999) The vanishing skylords. WWF-India Network *Newsletter* 9(4): 13-18.

Satheesan, S. M. (2000a) *Vultures in Asia.* Pp. 165-174 in R. D. Chancellor and B.-U. Meyburg. *Raptors at risk.* Proceedings of the Fifth World Conference on Birds of Prey and Owls. Surrey, UK: Hancock House Publishers.

Satheesan, S. M. (2000b) Vulture-eating communities in India. *Vulture News* 41: 15-17.

Satheesan, S. M. (2000c) The role of poisons in the Indian vulture population crash. *Vulture News* 42: 3-4.

Singh, R. B., Grubh, R. B. and Pandey, K. C. (1996) Movements of Indian White-backed Vultures *Gyps bengalensis* between their permanent feeding ground and roosting sites near Bombay, India. *Vulture News* 34: 3-7.

St John, O. B. (1889) On the birds of southern Afghanistan and Kelat. *This* (6)1: 145-180.

Thiollay, J. M. (2000) Vultures in India. *Vulture News* 42: 36-38.

Trivedi B.P. India Vulture Die-Offs spurs carcass crisis *National Geographic Today, Dec 2001.*

Vijayan, V. S. (1991) Keoladeo National Park Ecology Study 1980-1990. Final report. Bombay Natural History Society.

Virani, M., Gilbert, M., Watson, R. T., Oaks, J. L., Benson, P. C. and Khan, A. A. (2002) Aspects of breeding and mortality of the Oriental White-rumped Vulture *Gyps bengalensis* in the Punjab Province, Pakistan. Unpublished ms.

Virani, M., M. Gilbert, R. Watson. L. Oaks, and P. Benson, A.A. Khan, and H. S. Baral. (2001) Asian vulture crisis project: field results from Pakistan and Nepal for the 2000-2001 field season. Reports from the workshop on Indian Gyps vultures. Pages 7-9 in t. Katzner and J. parry-Jones, editors. Proceedings of the 4th Eurasian congress on raptors. Estacion Biologica Donana, Raptor research foundation, Seville.

Virani, M., Vulture colony in the southwestern Cape Province. MSc dissertation,Vulture Gyps coprotheres. J. Anim. Ecol. 50:815-825.Vulture Gyps corprotheres. Journal ofAnimal Ecology 50: 815-82.

Virani, M., Watson, R., Risebrough, B. & Oaks, L (2000). Should Africa brace itself for an imminen vulture epidemic? Lessons from the Asian vulture crisis. *Oral presentation at the Pan-Africa Ornithological Congress. Kampala, Uganda.*virus. Bull. Off. int. Epiz. 18: 118-148.

Wyk, E. V., Bank, F. H. V. D. and Verdoom, G. H. (1993) The Cape Griffon *Gyps coprotheres*-a conservation priority. *Vulture News* 29: 4-18.

Wynne-Edwards, V. C. (1955) Low reproductive rates in birds, especially Sea birds. *Acta* X1. International Ornithological Congress 1954 pp. 540-547.

1396640R0

Printed in Great Britain by
Amazon.co.uk, Ltd.,
Marston Gate.